The Instrumental Spectrometric and Spectroscopic Analysis of Pheromones

A Handbook

Dr. Nick Winstone-Cooper

and

Jasmine Tripconey

First published by Ellixia Publishing Limited in 2021 (www.ellixia.com)

978-1-008-96278-1

Acknowledgements: Most images and diagrams have been created by the writers and those from other sources are acknowledged with thanks for the permission to use them.

Written for

and the Welsh Government's educational Hwb.

Dr. Nick Winstone-Cooper studied chemistry and physics at Cardiff University before completing postgraduate research in nuclear chemistry, focusing on the creation of radioactive complexes of macrocyclic phosphines for application as heart and bone imaging agents in cancer diagnosis.

He worked extensively in the United Kingdom, France, North America, Italy and the Republic of Korea before moving into education and education consultancy.

Jasmine Tripconey studied Chemistry and Drug Discovery at the University of Bath. Her interests include the biochemistry of selective serotonin reuptake inhibitors and bioinorganic chemistry.

Table of Contents

Introduction

Whilst the title of this volumes refers to *pheromones* which is a well known term, they comprise a sub-class of *semiochemicals* which comprise two classes of signalling chemicals that birds, fish and mammals can detect in their environment and which may modify their behaviour. The other sub-class of semiochemicals are *allelochemics* which are described, briefly, before the focus of this text changes to the subject of this text, pheromones. Any dog walker will observe that the animal will follow a trail sniffing the ground as s/he proceeds whilst the odour is rarely apparent to the dog walker. This is because dogs have many thousand more odour receptors in their noses than do humans. The odour receptors do not duplicate themselves and are respond to different odours hence dogs are much more sensitive to the environment than are humans even if inhabiting the same space. This also explains why some dogs are so effective at detecting drugs, dead bodies and even explosives.

The ability to detect odours and trails undetectable by humans also explain how bees and ants behave. It has been reported recently that bees can even detect and identify the location of explosives such as trinitrotoluene (TNT) and ants follow land based trails which inform their behaviour and lead to collaboration between the inhabitants of the nest. It is important, however, to remember, though, that not all odours are pheromones and not all pheromones have an odour as, for example, they can be communicated in water or as land laid trails.

That animals can detect, by smell, substances undetectable by humans has been long suggested. For example, in 1623, Charles Butler wrote in *The Feminine Monarchie* that an injured bee's *'ranke smell'* attracts bees and being angry will sting humans whilst, in the late 19th century, Joseph Lintner recorded that huge numbers of male silk moths would congregate around a female moth. Lintner concluded that they were attracted by some sort of odour in other words, they detected a chemical substance produced by the female moth.

It was only in the twentieth century, however, that pheromones have been isolated and characterised. The first pheromone was identified by Adolf Butenandt and for this work he shared the 1939 Nobel Prize in chemistry with Leopold Ružička. Butenandt was unable to accept the award though since, in 1937, Hitler had banned Germans from accepting the award after Carl von Ossietzky was awarded the 1936 Nobel Peace Prize. He did finally receive the medal, but not the prize money, in 1949.

Allelochemics

Allelochemics (a term suggested by Whittaker in 1970 describes chemicals that mediate interspecific interactions and there are several sub-categories:

- *Allomones*: which benefit the emitting species such as venom secreted by wasps.
- *Kairomones*: which instructs the recipient perhaps by identifying locations.
- *Synomones*: Those which mediate interactions and benefit both the recipient and the emitting species.

These molecules are not considered in this text as the focus is on pheromones.

Pheromones

A *pheromone*, a term coined in 1959 by Karlson and Lüscher, refers to any substance or group of substances emitted by any organism to which another organism will receive and respond.

There are a number of categories of pheromones and they are as follows:

- **Sex pheromones** are chemicals that primarily affect an interaction between the sexes however many types of bacteria also release pheromones to attract other cells to agglomerate and absorb DNA from other cells which is incorporated into their own DNA, a process known as transformation. This is one way in which yeast spreads and mould develops.

  Butterflies and bees also release pheromones to attract a mate. These pheromones must be volatile as they can travel up to ten miles. They also release pheromones to repel competitive creatures. Through their urines, boars and pigs spray pheromones into the sty and indicates the sows who are ready to be fertilised whilst sea urchins are known to release pheromones into the surrounding water which sends a chemical message to other sea urchins to release their eggs. This leads to the entire colony releasing their eggs at the same time.

  An example of a sex pheromone is *bombykol* which is released by the female silkworm moth and is the first pheromone to have been full isolated and characterised.

- **Aggregation pheromones** are molecules which increase the population density of the animals (often both sexes) in the vicinity of the pheromone. Most pheromones are produced by females and have been found to be of importance in butterflies. Commercially, they are used to control weevils in flour.

- *Alarm pheromones* are stimulate escape or defence behaviour. When attacked, many species release a volatile substance which can trigger flight (as with aphids) or aggressive defence and the latter has been observed with ants and bees. The phenomenon has also been observed in some plants especially trees. When cut, trees and other plants can release a pheromone along a line of the same species leading their fellow plants to take actions to protect themselves. Alarm pheromones are typically comprised of a complex mixture of esters and alcohols and may contain as many as forty different chemical substances although some are simple mixtures of a few chemicals. For example, the mixture released by stinging bees includes 3-methylbutyl ethanoate, butyl ethanoate and pentan-1-ol. When a beekeeper is collecting honey from the beehive, a smoke generating device is used to mask the smell and calm the bees.

- *Dispersal pheromones* advise creatures to flee. Few have been fully characterised.

- *Epideitic phermones* are typically deposited by egg layers to suggest to others of their species that they should find somewhere else to lay their own eggs.

- *Releaser pheromones* stimulate females to ovulate and male creatures to produce testosterone. They act to, perhaps, to attract a mate. These pheromones are very volatile and can travel several miles but degrade very quickly in the environment.

- *Maturation pheromones* speed up the sexual maturation of creatures and often lead to maturation in huge numbers of the swarm. This is one reason why locusts can suddenly appear and swarm in tens of millions as has recently occurred (March 2021) in North Eastern Africa. Every day, locusts can consume their body mass in feeding and can strip a field of its crops in minutes.

- *Territorial pheromones* are boundary markers which are best known for being produced in the urine of cats and dogs. Interestingly, some seabirds also deposit pheromones to mark their nests and territorial boundaries instructing or encouraging other members of their species to find somewhere else to nest.

- *Trail pheromones* are secreted by workers of social insects to recruit other individuals to food sources or to a new colony site. Common in ants who mark their paths to enable other members of the species to forage food, their repeated visits leads to continual renewal of the trail and when they stop visiting, because the food is exhausted, the trail evaporates because the volatile pheromones are no longer being replaced. Examples also wasps who, having found a new nesting site, lay a trail for other wasps to find it and caterpillars who create a trail for group travel.

■ **Worker pheromones** are a large class of behaviour controlling pheromones which encompass chemicals which instruct worker bees and ants as to their role but also can act to calm a herd of animals. There is another subclass, *necromones*, which are given off by a decomposing dead organism which enable crustaceans to identify a source of food.

Much of the work on pheromones has been conducted on bees and ants. Neither creature has a nose. Ants detect chemical substances with their antennae and follow trails whilst bees detect substances with their antennae, mouths and the tips of their legs and detect these substances even when flying. This is fascinating since, for creatures with noses, it has been estimated dogs have five to eight times the number of odour receptors in the nose than humans. In contrast, bees, especially honey bees, have a fifty to one hundred times higher ability to detect pheromones than do dogs.

Queen and Worker Bees

There has been much work studying the interactions of bees and ants in their respective colonies not least to determine how a Queen Bee can control the nest and why ants, especially worker and farmer ants can work together to build their *cities* and harvest food. It has been concluded that they work together due to *signalling pheromones*.

It appears that the Queen bee produces a *Queen pheromone* when fertile and laying eggs which instructs the worker bees to not lay eggs. Since a typical beehive can contain half a million bees a chemical trail is the only way to pass on the instruction.

If the Queen is removed then there is no longer any message and the worker bees then begin to lay their eggs. In many cases the messenger pheromone is a, non-volatile, saturated long-chain hydrocarbon molecule and some of these are analysed in this wolume.

Honey bees (*Apis mellifera*) possess one of the most complex systems of pheromone communication systems yet discovered. They possess at least fifteen glands producing a huge array of compounds including one which produces a pheromone of at least forty different compounds. They are produced and received by Queen bees, drones and worker bees and either elicit a response or respond to a chemical messenger from other bees.

Any of the eggs can develop, as larvae, in to Queen bees, and that is controlled by the excretion of Royal jelly[1]. If only Royal jelly is fed the larva will develop into a Queen bee and this is caused by the pheromones in the Royal jelly. If not fed Royal jelly then the larvae turn into worker bees. Queen bees, which can live for up to two years can produce one million eggs whilst the worker bees rarely live beyond a couple of months.

Two *alarm pheromones* have been definitely identified in worker honey bees. One consists of at least forty different compounds including alcohols and esters. Being very volatile and, with the esters readily degraded by water, they are not particularly specific and do not last very long. This may not matter since alarm pheromones are only released when a bee stings another creature and alerts local bees to the scene so its inability to travel far is insignificant. The other alarm pheromone, released from the mandibular glands, is essentially 2-heptanone. Also highly volatile it has a repellent effect and bees use it as an anaesthetic to paralyse intruders. After the intruders have been paralysed the bees remove them from the hive.

Worker bees also produce *Nasonov pheromones* which have a role in orientation and recruitment. Three of these pheromones are analysed later in this volume but we must also be aware of the *Queen Retinue Pheromone* which is a collection of compounds which are used to attract worker bees around the queen bee. One of these, cetyl alcohol is analysed later in this volume.

Human Uses of Pheromones

It has long been recognised that pheromones can be used to control insects and there have been periodic suggestions that human beings also produce pheromones. No human pheromones have been conclusively identified but one of the pheromones analysed in this volume is used in a commercial male fragrance.

There are, however, other practical uses and it has been reported recently (February 2021) that the National Trust in England are planning to experiment with pheromones at Blickling Hall, reportedly the birthplace of Anne Boleyn, to protect the tapestries, curtains and clothes from an influx of moths using small pieces of cardboard coated with the pheromones. The idea is that, attracted by the scent, the male moths will be drawn to it and get coated with the powder and become confused. This means that the eggs laid by females go unfertilized. That, at least, is the theory.

---

[1] Typically, Royal jelly comprises 66% water, 10-15% protein, 10-15% sugars, 6-10% fatty acids and 2-5% 10-hydroxy-2-decenoic acid together with trace amounts of minerals, antibacterial and antibiotic compounds as well as vitamins B5, B6 and C.

Pheromones are fascinating from several perspectives:

- Apparently simple molecules have a huge effect on an environment and huge numbers of species whilst largely unnoticed by human beings.

- Some of these molecules, whilst simple in structure are actually quite tricky and time – consuming to synthesise in reasonable yields in the laboratory but creatures create them without even knowing it.

- The environments experienced by insects, dogs and cats even when in the same geographic location are all very different to those that humans experience and the human perspective of the world is simply one experience of the environment we all occupy.

In this volume we investigate and determine the structure of twenty different pheromones.

Each pheromone is introduced by its trivial name and its use in nature discussed. The empirical composition and formula mass (Mr) are used to state the empirical and molecular formulas. In each chapter we then investigate the infrared and mass spectra to draw elementary suggestions about the structure of the molecule. We then analyse and fully interpret the ^{1}H and ^{13}C nmr spectra to draw a final conclusion about the structure of the molecule. Each chapter concludes with display of the concluded structure and the molecule's systematic name.

Part I

Elemental data and correlation charts

The following pages present:-

- A truncated Periodic Table showing the first four periods of the elements; which include all those elements involved in the compounds discussed in this series;

- Correlation charts for ^{1}H and ^{13}C nmr spectroscopy;

- A correlation chart for the analysis of infra red spectra;

- A concise summary of assignable fragments in the electron ionisation (EI) mass spectrometry analysis.

The Periodic Table of the Elements

(1)	(2)												(3)	(4)	(5)	(6)	(7)	(0)
1 **H** hydrogen 1.0																		**18** 2 **He** helium 4.0

Key
atomic number
Symbol
name
relative atomic mass

Group																		
3 **Li** lithium 6.9	4 **Be** beryllium 9.0												5 **B** boron 10.8	6 **C** carbon 12.0	7 **N** nitrogen 14.0	8 **O** oxygen 16.0	9 **F** fluorine 19.0	10 **Ne** neon 20.2
11 **Na** sodium 23.0	12 **Mg** magnesium 24.3	**3**	**4**	**5**	**6**	**7**	**8**	**9**	**10**	**11**	**12**		13 **Al** aluminium 27.0	14 **Si** silicon 28.1	15 **P** phosphorus 31.0	16 **S** sulfur 32.1	17 **Cl** chlorine 35.5	18 **Ar** argon 39.9
19 **K** potassium 39.1	20 **Ca** calcium 40.1	21 **Sc** scandium 45.0	22 **Ti** titanium 47.9	23 **V** vanadium 50.9	24 **Cr** chromium 52.0	25 **Mn** manganese 54.9	26 **Fe** iron 55.8	27 **Co** cobalt 58.9	28 **Ni** nickel 58.7	29 **Cu** copper 63.5	30 **Zn** zinc 65.4		31 **Ga** gallium 69.7	32 **Ge** germanium 72.6	33 **As** arsenic 74.9	34 **Se** selenium 79.0	35 **Br** bromine 79.9	36 **Kr** krypton 83.8

^{1}H and ^{13}C NMR Correlation Charts

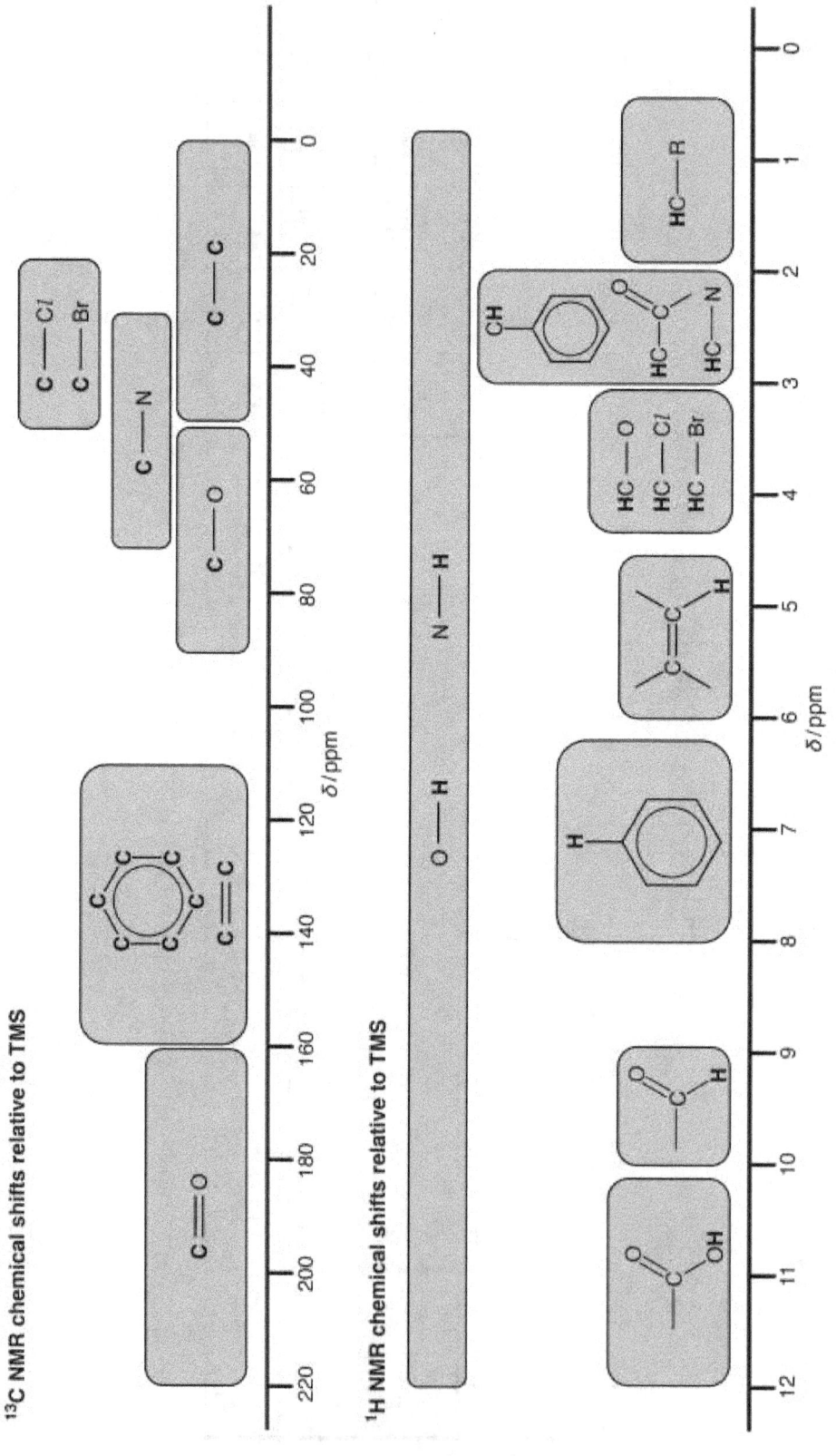

Infrared Correlation Chart

Bond	Functional group	Wavenumber (cm⁻¹)
Aromatic C = C	Aromatic compounds	1450 – 1650 (Multiple peaks)
C ≡ N	Nitriles	~ 2250
C – H	Alkyl groups, alkenes and aromatic compounds	2850 – 3000 (alkanes) 3000 – 3200 (alkenes and aromatics)
O – H	Carboxylic acids	2500 – 3500
N – H	Amines and amides	3300 – 3500
O – H	Alcholos and phenols	3200 – 3600

Bond	Functional group	Wavenumber (cm⁻¹)
C – C	Alkanes and alkyl groups	750 – 1100
C – X	Haloalkanes (X = Cl, Br or I)	500 – 800
C – F	Fluoroalkanes	1000 – 1350
C – O	Alcohols, carboxylic acids and esters	1000 – 1300
Aliphatic C = C	Alkenes	~ 1650
C = O	Aldehydes, ketones, carboxylic acids, esters and acid chlorides	~ 1750

Chart axis (cm⁻¹): 3600, 3200, 3000, 2800, 2600, 2400, 2200, 2000, 1800, 1600, 1400, 1200, 1000, 800, 600, 400

Boxes: RO – H; N – H; ArC – H; – C – H; = C – H; RCOO – H (carboxylic acids); C ≡ N; C = O; C = C; Ar – H; C – O; C – C; C – X

Notes: In the table above, Ar refers to an aromatic ring such as benzene whilst X refers to any of F, Cl, Br or I.

Some peaks are of little use for identification purposes since, for example, most organic compounds contain C – H bonds and so the presence of peaks just below 3000 cm⁻¹ is of little use for identification purposes. There is, however, a distinction between the C – H peaks above and below this wavenumber: aromatic compound C – H bonds appear above i.e. to the left of 3000 cm⁻¹ whilst aliphatic C – H bonds appear below, to the right, of 3000 cm⁻¹.

The combination of peaks is also important. For example, a carboxylic acid contains both a C =O and O – H bond so the presence of both is necessary for the confirmation of this class of compound.

Common Mass Fragmentation Ions

The following table shows a large number of fragmentation assignments which are relevant to the molecules in this volume. There are a number of important points to note:-

- If a molecule contains a *chlorine* atom then that fragment will exhibit two peaks, two units apart, due to the existence of the ^{35}Cl and ^{37}Cl isotopes. The heights of these peaks will be in the proportion 3:1 and there will always be two other peaks at m/z = 35 and 37, also in the ratio 3:1 due to the natural occurred of the isotopes. If any of these four peaks are absent then chlorine is not present. The presence of chlorine will already, however, be recorded in the elemental composition of the compound.

- Similarly to chlorine, if a molecule contains a *bromine* atom then there will be two peaks for the molecular fragment due to the existence of ^{79}Br and ^{81}Br. Since these two isotopes exist in nearly equal proportions then the peak heights of these fragments will be of approximately equal height and there will, of course, also be peaks of equal height at m/z 79 and 81.

- *Aromatic* compounds such as those containing a benzene ring will usually exhibit a peak at m/z = 77 due to the existence of the C_6H_5 – functional group. If there is a peak at this m/z ratio then it is almost always due to this. There will then also be peaks of lower m/z value due to fragmentation of the ring but these can also be due to the fragmentation of aliphatic chains and so the clue is in the m/z = 77 peak. More highly substituted benzene rings will exhibit peaks at m/z = 76, 75 etc; The existence of the aromatic portion of the molecule is, however, also and *always* conclusively demonstrated by the ^1H and ^{13}C nmr spectra. In the following table, the aromatic fragments are indicated by [a].

- Many mass spectrometry measurements may be completed in little more than one second. Whilst in everyday life one second is very brief it is, in physical terms, quite long. This means that the unstable ionised molecule may fragment or rearrange itself on its journey along the apparatus resulting in peaks that would not be predicted simply by considering the ripping apart of a molecule. peaks in the table below which result from rearrangement of an ionised molecule or fragments are indicated by [b] after the fragment's formula.

- Some molecules may produce different fragments of the same m/z and are listed as bullet points in the table below.

Table of mass fragments

m/z	Ion	m/z	Ion
15	$[CH_3]^+$	65	$[C_5H_5]^{+a}$
17	$[OH]^+$	67	$[C_5H_7]^+$
18	$[H_2O]^+$	69	$[C_5H_9]^+$
26	$[CN]^+$	70	$[C_5H_{10}]^+$
27	$[C_2H_3]^+$	71	• $[C_5H_{11}]^+$ • $[C_3H_7\text{-}C{=}O]^+$
28	$[C_2H_4]^+$	72	$[C_2H_5\text{-}CO\text{-}CH_2{+}H]^{+b}$
29	$[C_2H_5]^+$ $[CHO]^+$	73	• $[C_3H_7OCH_2]^+$ • $[C_2H_5O\text{-}C{=}O]^+$ • $[C_3H_7CHOH]^+$ • $[C_2H_5OCHCH_3]^+$
30	$[CH_2NH_2]^+$	74	$[CH_2\text{-}COOCH_3{+}H]^{+b}$
31	• $[CH_2OH]^+$ • $[OCH_3]^+$	75	• $[C_2H_5O\text{-}C{=}O{+}2H]^{+b}$ • $[C_2H_5COO{+}2H]^{+b}$
35 & 37	$[^{35}Cl]^+$ & $[^{37}Cl]^+$	77	$[C_6H_5]^{+a}$
39	$[C_3H_3]^{+a}$	78	$[C_6H_5{+}H]^{+ab}$
40	$[CH_2CN]^+$	79	• $[C_6H_5{+}2H]^{+ab}$ • $[^{79}Br]^+$
41	• $[C_3H_5]^+$ • $[CH_2CN{+}H]^{+b}$	81	• $[C_6H_9]^+$ • $[^{81}Br]^+$
42	$[C_3H_6]^+$	82	• $[C_6H_{10}]^+$ • $[C^{35}Cl^{35}Cl]^+$
43	$[C_3H_7]^+$ $[CH_3C{=}O]^+$	83	• $[C_6H_{11}]^+$ • $[CHCl_2]^+$ (also 85&87)
44	$[CH_3CH\text{-}NH_2]^+$	84	• $[C_6H_{12}]^+$ • $[C^{35}Cl^{37}Cl]^+$
45	• $[CH_3CHOH]^+$ • $[CH_2OCH_3]^+$ • $[CH_2CH_2OH]^+$ • $[COOH]^+$	85	$[C_6H_{13}]^+$ $[C_4H_9\text{-}C{=}O]^+$
49	$[CH_2{}^{35}Cl]^+$	86	• $[C_3H_7\text{-}CO\text{-}CH_2{+}H]^{+b}$ • $[C^{37}Cl^{37}Cl]^+$
50	$[C_4H_2]^{+a}$	88	• $[CH_2\text{-}COOC_2H_5{+}H]^{+b}$
51	$[CH_2{}^{37}Cl]^+$ $[C_4H_3]^{+a}$	89	• $[C_3H_7\text{-}O\text{-}C{=}O{+}2H]^{+b}$ • $[C_3H_7COO{+}2H]^{+b}$
52	$[C_4H_4]^{+a}$	90	• $[C_6H_5\text{-}CH]^+$
53	$[C_4H_5]^+$	91	• $[C_6H_5\text{-}CH_2]^+$ • $[C_6H_5\text{-}CH{+}H]^{+b}$
54	$[H_2CH_2CN]^+$ $[CH_3CHCN]^+$	92	• $[C_6H_5\text{-}CH_2{+}H]^{+b}$

m/z	Ion	m/z	Ion
55	$[C_4H_7]^+$	93	$[C_7H_9]^+$ $[CH_2{}^{79}Br]^+$
56	$[C_4H_8]^+$	94	$[C_6H_5O+H]^+$
57	$[C_4H_9]^+$ $[C_2H_5\text{-}C\text{=}O]^+$	95	$[CH_2{}^{81}Br]^+$
58	$[CH_3\text{-}CO\text{-}CH_2+H]^{b}$	97	$[C_7H_{13}]^+$
59	$[C_2H_5OCH_2]^+$ $[CH_3O\text{-}C\text{=}O]^+$ $[C_2H_5CHOH]^+$ $[CH_3O\text{-}CHCH_3]^+$	105	$[C_6H_5C\text{=}O]^+$ $[C_6H_5\text{-}CH_2CH_2]^+$
60	$[CH_2\text{-}COOH+H]_+{}^{b}$	107	$[C_6H_5\text{-}CH_2O]^+$
61	$[CH_3COO+2H]_+{}^{b}$ $[CH_3OCO+2H]_+{}^{b}$	108	$[C_6H_5\text{-}CH_2O+H]_+{}^{b}$
63	$[C_5H_3]^{+a}$	119	$[C_6H_5\text{-}C(CH_3)_2]^+$

[a] Good diagnostics for benzene ring compounds

[b] Where a fragment results from a rearrangement with the movement of *one* hydrogen atom from one carbon to another this is indicated by the fragment followed by +H.

^{1}H NMR Coupling Constants

Aliphatic Alkenes

Isomerism	Coupling constant (J) range (Hz)
Geminal	0 – 5
Vicinal (cis) / Vicinal (Z –)	5 – 14
Vicinal (trans) / Vicinal (E –)	15 – 20

Substituted Aromatic Compounds

Designation	Ortho –	Meta –	Para –
Structural formula			
Coupling constant range (J):	7 – 10	2 – 3	0 – 2

Part II

The Analysis of Pheromones

This section comprises twenty chapters which investigate the application of infrared spectroscopy, mass spectrometry and multinuclear magnetic spectroscopy (^{1}H and ^{13}C) to the determination of the structures of a wide range of pheromones.

These range from bee, moth and locust pheromones through to those employed by rabbits and wolves.

Chapter 1

Guaiacol

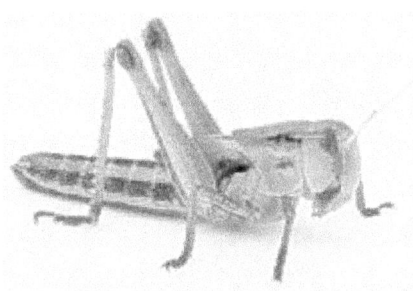

Guaiacol is a commonly found substance, occurring in celery seeds, tobacco, orange leaves and lemon peel. It is used by locusts when swarming in order to direct the swarm. This is particularly topical at the time of writing (May 2021) as large parts Ethiopia and Kenya have had their crops devoured by swarms of tens of millions of locusts which can eat their entire body weight in food every day.

At room temperature this substance is a crystalline solid but, with a melting point of just 26°C, it quickly becomes a colourless oil. It has a relatively high melting point of 205°C which indicates that there are significant intermolecular interactions holding the molecules together and it is relatively soluble in water (25g dm^{-3} at 25°C)

This compound has the **elemental composition**: C: 67.70%, H: 6.52%, O: 25.80% and has the **formula mass** (M_r) of 124.08 g mol^{-1}.

This means that the empirical and molecular formulas are both $C_7H_8O_2$.
The presence of two oxygen atoms implies the existence of either a carboxylic acid or an ester and the existence of more than six carbon atoms also suggest that the molecule contains a long, aliphatic, chain or a benzene ring.

Guaiacol

Infrared Spectrum

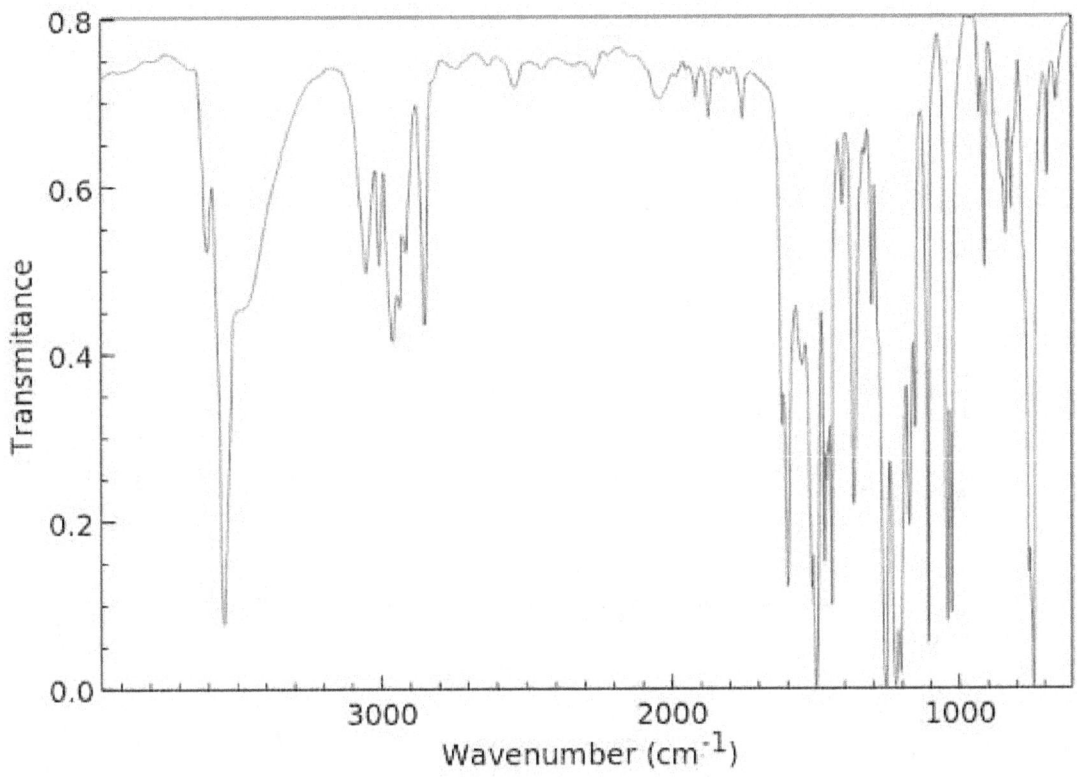

Observations

(√ / X)	Wavenumber range (cm⁻¹)	Wavenumber (cm⁻¹)	Assignment
√	3200 - 3700	3550	**O – H**
X	3200 - 3600		**N – H**
√	3000 – 3300	3040	**C – H (aromatic)**
√	2500 – 3000	2800 – 3000 (multiple)	**C – H (aliphatic)**
X	2200 – 2500		**C ≡ N**
X	1700 – 1800		**C = O**
X	1600 – 1700		**C = C (aliphatic)**
X	1585 – 1600		**C – C (aromatic)**
X	1450 – 1600		**C – C (aromatic)**
X	1000 – 1300	1250	**C – O**
X	700 – 1000		**C – X** (X = Cl, Br or I)

Conclusions

This compound is aromatic and possesses an – OH group. The peak at 1250 cm⁻¹ also indicates the existence of an alkoxy (O – C) group.

Guaiacol

Mass Spectrum

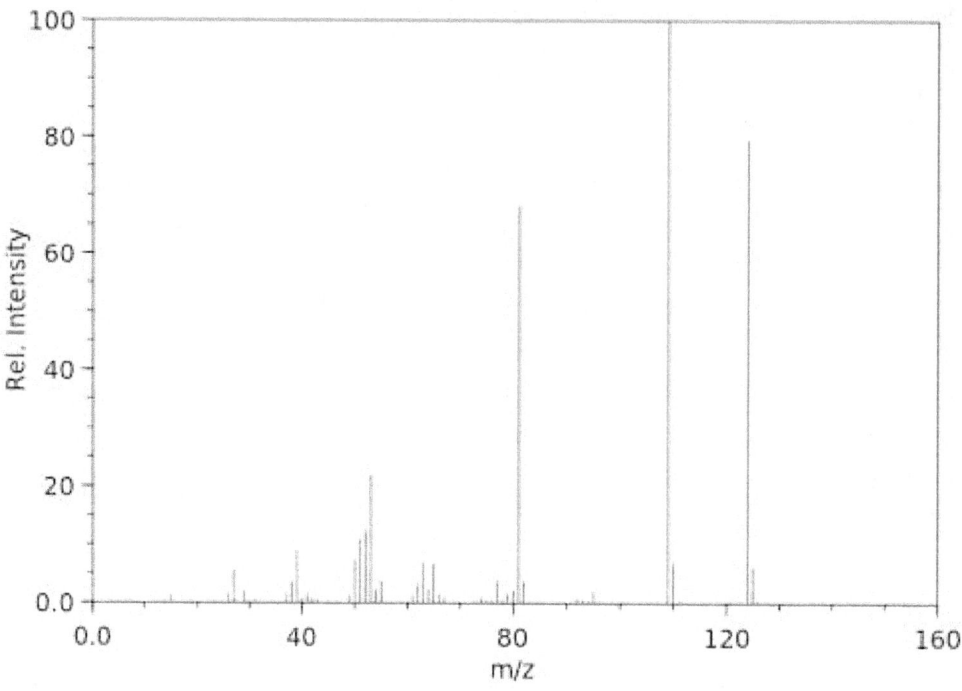

Observations

Charged fragments (m/z)	Assignment	Charged fragments (m/z)	Assignment
Molecular ion: 124	$[C_7H_8O_2]^+$	Base peak: 109	$[C_6H_5O_2]^+$
109	$[C_6H_5\text{-}CH_3O]^+$	52	$[C_4H_4]^+$
81	$[C_6H_9]^+$	51	$[C_4H_3]^+$
63	$[C_5H_3]^+$	39	$[C_3H_3]^+$
65	$[C_5H_5]^+$	29	$[CHO]^+$
53	$[C_4H_5]^+$	15	$[CH_3]^+$

Conclusions

The multiple peaks with 4,5,6 and 7 carbon atoms implies the presence of a benzene ring and the molecular ion and base peaks implies this molecule has one of the following structures:

NMR Spectra

The three potential candidates will produce similar ^{1}H nmr spectra but there will be significant differences in the predicted spectrum:

- All will produce a signal, of integral one, due to the – OH group.

- There will be a singlet of integral three, assignable to the – OCH$_3$ group but no splitting due to the –CH$_3$ being bonded to an oxygen atom so no adjacent hydrogen atoms.

- There will be peaks in the aromatic region ($\delta 6.2$ – 8ppm) with a total integral of four but each will produce a different pattern of multiplets. We discuss these in turn referencing the hydrogen atoms by the letters as shown:-

- **Candidate I**
  Predictions

On its own with no hydrogen atoms on adjacent carbon atoms, H(a) would produce a singlet but in this case it will be split into a doublet by H(c) which, in turn, would be split into a doublet of doublets by H(c) which would be split into a doublet of doublets of doublets by H(d) as shown below:

H(a) is split into a doublet by H(b)

This doublet is split into a doublet of doublets by H(c)

And this is split into a doublet of doublets of doublets by H(d)

We can go one step further, though, by considering the coupling constants.

- H(a) is ortho- to H(b), meta- to H(c) and para- to H(d);
- H(b) is ortho- to H(a) and H(c) and para- to H(d);
- H(c) is ortho- to H(b) and H(d) and meta- to H(a);
- H(d) is ortho- to H(c), meta- to H(b) and para- to H(a)

This means that we should observe coupling constants due to ortho-, meta- and para- locations in the aromatic multiplets. They may overlap but the para- coupling constant should certainly be clear.

Candidate II

Predictions

- H(e) will split into a doublet by H(f) and split into a doublet of doublets by H(g) and then into a doublet of doublets of doublets by H(h);

- H(f) will be split into a doublet by H(e) and the doublet will be split into a doublet of doublets by H(g). Equally, we can work the other way in that H(f) will be split into a doublet by H(g) and the doublet will be split into a doublet of doublets by H(g);

- This doublet of doublets will be split into a doublet of doublets of doublets by H(h).

H(f) is split into a doublet of doublets by H(e)

This doublet is split into a doublet of doublets by H(g)

This doublet of doublets is split into a doublet of doublets of doublets by H(h)

With regard to coupling constants:-

- H(e) is ortho- to H(f) and meta- to H(g) and H(h);
- H(f) is ortho- to both H(e) and H(h) and para- to H(h);
- H(g) is ortho- to H(f) and meta- to both H(e) and H(h);
- H(h) is is meta- to H(e) and H(g) and is para- to H(f).

This means that there will be multiplets with coupling constants assignable to ortho-, meta- and para- couplings although, again, the multiplets might overlap.

Candidate III

Predictions

This candidate will have the simplest ^{1}H nmr spectrum since the molecule is, apparently, symmetrical with H(i) and H(i') being mutually chemically and magnetically equivalent. Equally, H(j) and H(j') are also mutually chemically and magnetically equivalent.

Both pairs of hydrogen atoms are bonded to carbon atoms so will be deshielded to a similar extent and the multiplets may well overlap. If we look at the splitting, we will observe the following:

- H(i), which if were no hydrogen atoms on adjacent carbon atoms would be a singlet, is split into a doublet by H(j). It is split into a doublet of doublets by H(j') and this doublet of doublets is split into a doublet of doublets of doublets of doublets.
- H(j) would also be a singlet if H(i) was missing, is split into a doublet of doublet by the actual presence of H(i). This doublet is split into a doublet of doublets by H(j') and this multiplet is split into a doublet of doublets of doublets by H(i').

H(i) is split into a doublet by H(j)

This multiplet is split into a doublet of doublets of doublets by H(i')

This doublet is split into a doublet of doublets by H(j')

Since the molecule is symmetrical we would draw the same conclusions if working anti-clockwise from H(i') or from H(j) or H(j') and the peaks will overlap. Significantly there are no meta hydrogen atoms so we should not expect any multiplet with a coupling constant between J =2 – 3 Hz.

Considering the three candidates again, we can predict the following spectra which are shown in turn with the integral shown numerically above the multiplets

Guaiacol

^{1}H NMR Spectrum

The predicted ^{1}H nmr spectra are as follows:

Candidate I:

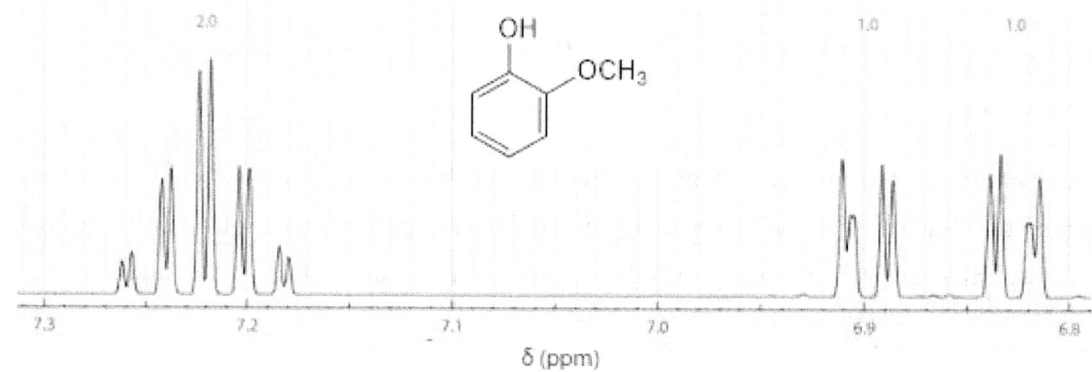

Candidate II:

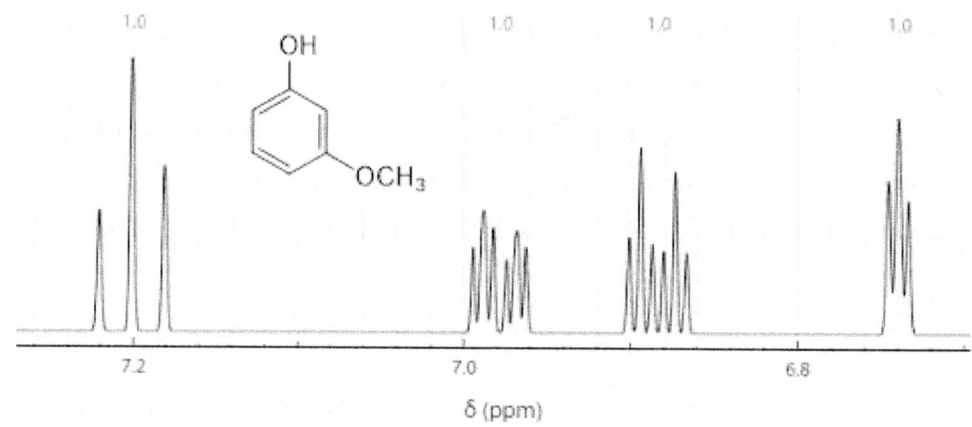

Candidate III:

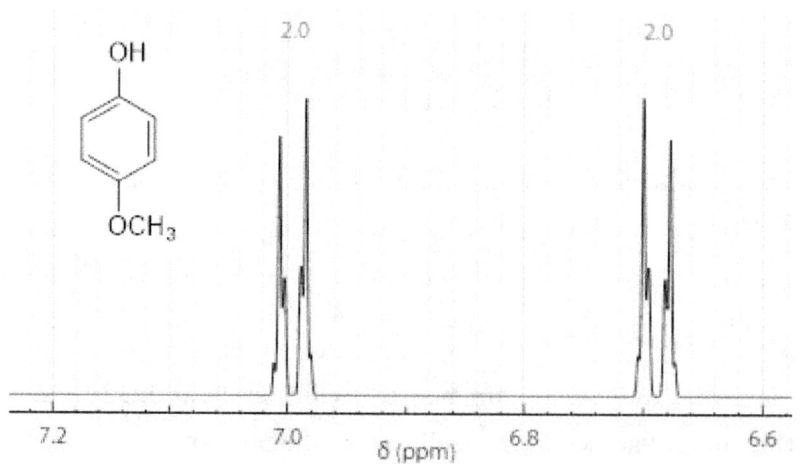

Guaiacol

The actual ^{1}H nmr spectrum is shown below:

▣ The peak at δ 5.75 ppm, of integral one, is clearly assignable to the phenolic hydrogen atom.

▣ The singlet, of integral three, at δ 3.9 ppm is assignable to a methoxy – OCH$_3$ group.

We need to examine the aromatic region in more detail:

δ (ppm)

Observations and Conclusions

Whilst the spectrum is complex, it becomes immediately clear that Candidate III can be discarded as it simply does not match the actual spectrum which, if this structure is correct should comprise two multiplets each of integral two which it does not.

This leaves us with the following possibilities:

I

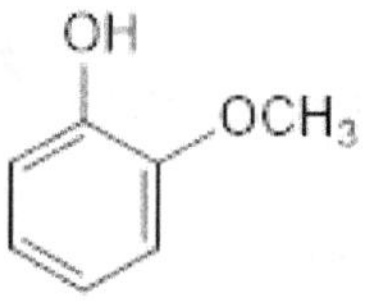

II

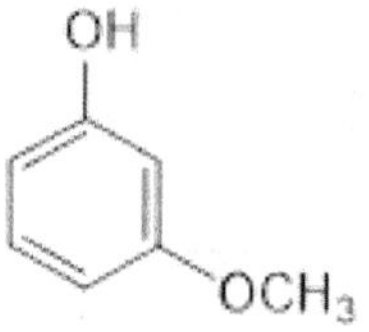

The observed spectrum contains two, closely spaced, multiplet with integrals in the ratio 2:1:1 and so the only possible structure is candidate I.

The coupling constants are of no use in this instance as both candidates contain ortho-, meta- and para- hydrogen atoms.

^{13}C NMR Spectrum

The ^{13}C nmr spectra of both candidates will be similar since both will contain a methoxy-carbon $(O - CH_3)$ peak in the range δ 50 – 90 ppm and six peaks due to the aromatic ring carbon atoms in the region δ 110 – 160 ppm.

If, however, we rotate the chosen candidate, we can observe that four of the carbon atoms are in two pairs of extremely similar chemical and magnetic environments. These are indicated by the hatched and solid rectangles in the structure below.

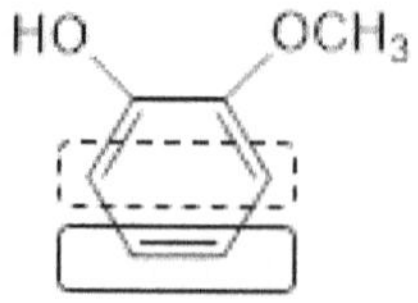

This would not occur with candidate II and there, indeed, are two pairs of peaks in the ^{13}C nmr spectrum.

Conclusions

Systematic name: 2 – methoxyphenol.

Chapter II

Methylparaben

Methylparaben, used a preservative in cosmetics and other personal care products, also acts as a pheromone for insects and especially bees and, surprisingly, wolves as well. The female wolves secrete the pheromone to deter wolves other than her mate from approaching when she is in heat. It also has some applications as a fungicide.

This substance has a melting point of 127°C and a boiling point of 275°C and so, at room temperature, it is a colourless crystalline solid. It is slightly soluble in water (2.5g dm^{-3}).

Methylparaben has the **elemental composition**: C: 63.12%, H: 5.31%, O: 31.56% and has the **formula mass** (M$_r$) of 152.15 g mol^{-1}.

These means that its empirical and molecular formulas are both $C_8H_8O_3$.

Methylparaben

Infrared Spectrum

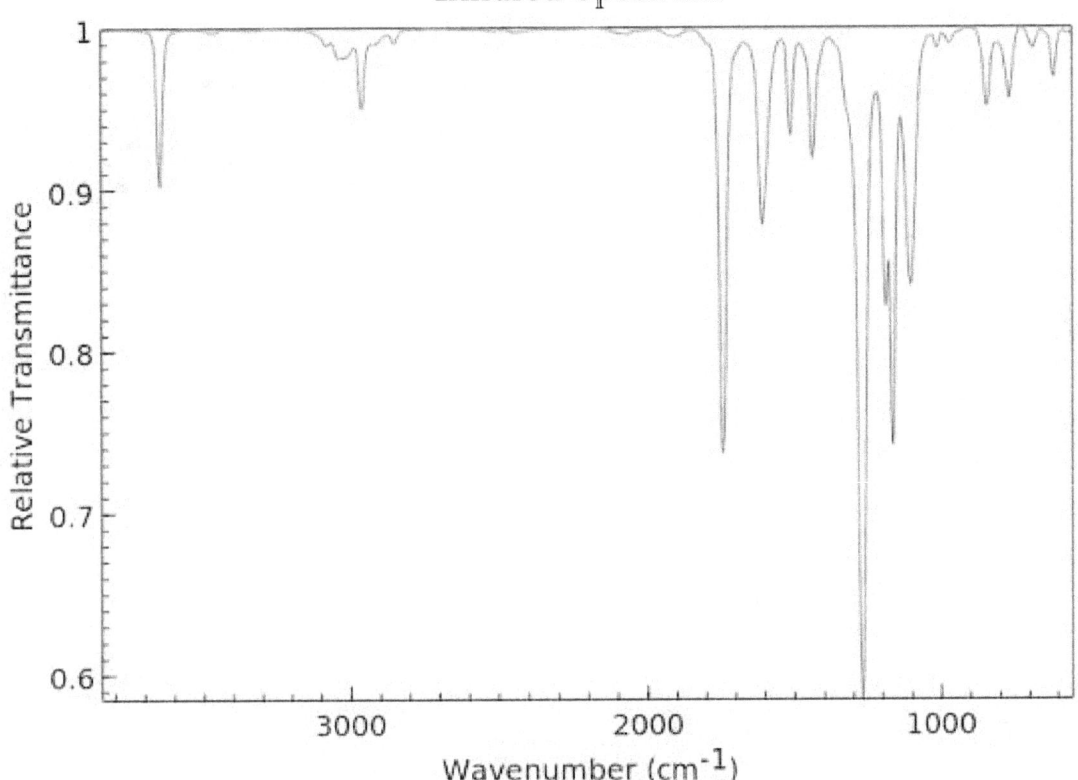

Observations

(√ / X)	Wavenumber range (cm⁻¹)	Wavenumber (cm⁻¹)	Assignment
√	3200 - 3700	3640	O – H
X	3200 - 3600		N – H
√	3000 – 3300	Broad weak peaks	C – H (aromatic)
√	2500 – 3000	Broad range	C – H (aliphatic)
X	2200 – 2500		C ≡ N
√	1700 – 1800	1740	C = O
X	1600 – 1700		C = C (aliphatic)
X	1585 – 1600	1585	C – C (aromatic)
X	1450 – 1600		C – C (aromatic)
√	1000 – 1300	1280	C – O
X	700 – 1000		C – X (X = Cl, Br or I)

Conclusions

This molecule is aromatic with aliphatic hydrocarbon groups; a C = O bond and a C – O

bond and so the molecule must be an aromatic ester probably with a – OH group on the ring.

Methylparaben

Mass Spectrum

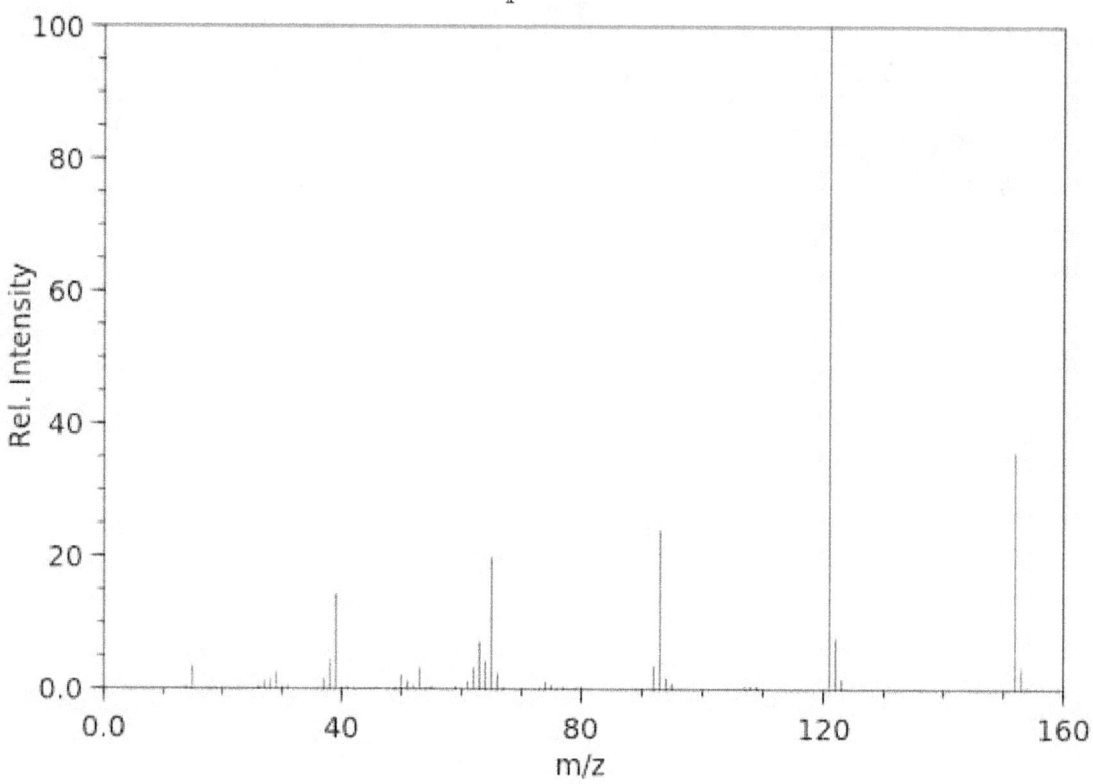

Observations

Charged fragments (m/z)	Assignment	Charged fragments (m/z)	Assignment
Molecular ion: 152	$[C_8H_8O_3]^+$	Base peak: 121	$[C_7H_5O_2]^+$

93	$[C_6H_5O]^+$	39	$[C_3H_3]^+$
65	$[C_6H_5]^+$	29	$[C_2H_5]^+$
43	$[C_2H_3O]^+$	15	$[CH_3]^+$

Conclusions

▪ The molecular ion is the ionised molecule and the significant difference is between the molecular ion and the base peak. The difference amounts to $[CH_3O]^+$ which implies this is a functional group or part of a functional group which has simply snapped off.

▪ The peak at m/z = 65 demonstrates the presence of a benzene ring with five hydrogen atoms attached.

▪ The peak at m/z = 43 implies the existence of a $- O - CH_3$ functional group which has also snapped off..

NMR Spectra

From the infrared spectrum we can conclude that the molecule contains an aromatic ring, a carbonyl group and a C – O bond. This immediately suggests that the compound has an ester functional group attached to a benzene ring.

The mass spectrum supports these assertions and the presence of the base peak assignable to $[C_7H_5O_2]^+$ leaves a group of m/z = 43 which can be assigned to $[C_2H_3O]^+$. This can be drawn as:

where the squiggle indicates an undetermined bond to, perhaps, the benzene ring.

This leads us to the concept that part of the structure is:

and this leads us to then place the other two oxygen atoms.

One must go on to the functional group otherwise the compound is not an ester and so we then have:

This leaves us with one more oxygen atom to place and, since we know from the infrared spectrum that the molecule has a phenolic – OH group we have three possible isomers as shown below:

and we consider each of these in turn starting with the ^{1}H nmr spectrum.

^{1}H NMR Spectra

Candidate I

This molecule can be drawn in two different ways as shown below but they are the same as flipping one drawing over produces the other one.

We will use the right hand structure and predict the spectral signals due to the hydrogen atoms working from the top left aromatic hydrogen atom.

Two signals will be easy to determine:

- The methoxy, $-OCH_3$, group will produce a singlet of integral three. It will be a singlet as there are no hydrogen atoms on the oxygen atom.
- The phenolic, $-OH$, group will produce a singlet of integral one. Hydroxyl groups can appear anywhere in the spectrum but will become self evident.

They will appear in the spectra of all three molecule and so are not helpful in the structure determination.

The important part of the spectrum are signals due to the four hydrogen atoms on the aromatic ring. Each will produce a signal of integral one but the signals will be split and might overlap and this is when coupling constants become of value. Alphabetically labelling the hydrogen atoms from the top left,

we can note that

H(a) is:-	H(b) is:-	H(c) is:-	H(d) is:-
■ ortho- to H(b)	■ ortho- to H(a)	■ ortho- to H(d)	■ ortho- to H(c)
■ meta- to H(c)	■ ortho- to H(c)	■ ortho- to H(b)	■ meta- to H(b)
■ para- to H(d)	■ meta- to H(d)	■ meta- to H(a)	■ para- to H(d)

and so we can expect coupling constants in all three of the ranges listed in the data table.

Methylparaben

If we look at the expected splitting of the signals we can predict the following:

■ The signal due to H(a) will be a doublet of a doublet of doublets as shown below:

1. H(a) is split into a doublet by H(b)

3. The doublet of doublets is split into a doublet of doublets of doublets by H(d)

2. The doublet is split into a doublet of doublets by H(c)

■ The signal due to H(b) will be a doublet of a doublet of doublets as shown below:

1. H(b) is split into a doublet by H(a)

3. The doublet of doublets is split into a doublet of doublets of doublets by H(d)

2. The doublet is split into a doublet of doublets by H(c)

We can work in other directions, from H(b) to H(c) and then to H(d) and then to H(a), for example but the result is the same: a doublet of doublets of doublets.

■ The signal due to H(c) will be a doublet of a doublet of doublets as shown below:

3. The doublet of doublets is split into a doublet of doublets of doublets by H(d)

2. The doublet is split into a doublet of doublets by H(a)

1. H(c) is split into a doublet by H(b)

■ The signal due to H(d) will also be a doublet of a doublet of doublets as shown below:

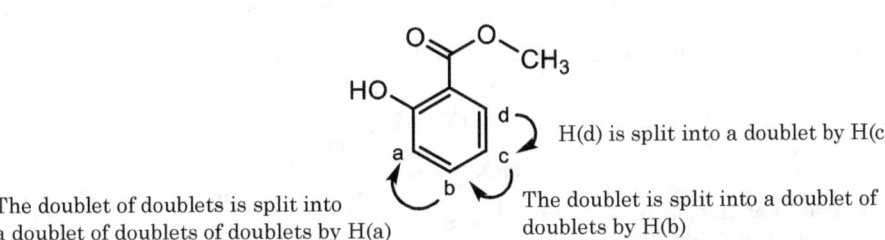

H(d) is split into a doublet by H(c)

The doublet of doublets is split into a doublet of doublets of doublets by H(a)

The doublet is split into a doublet of doublets by H(b)

Methylparaben

Candidate II

Using the same lettering we can predict the following multiplets:

- H(a) will be split into a doublet of doublets of doublets as detailed below:

1. H(a) is split into a doublet by H(b)

3. The doublet of doublets is split into a doublet of doublets of doublets by

2. The doublet is split into a doublet of doublets by H(c)

- H(b) will be split into a doublet of doublets by H(a) or by H(c). The resultant doublet will be split into a doublet of doublets by the other adjacent hydrogen atom and this doublet of doublets will be split into a doublet of doublets of doublets by H(d). The diagram below assumes that the splitting starts with H(a) but it is equally valid if the first splitting is by H(c).

1. H(b) is split into a doublet by H(a)

3. The doublet of doublets is split into a doublet of doublets of doublets by

2. The doublet is split into a doublet of doublets by H(c)

- H(c) will also produce a doublet of doublet of doublets as shown below:

3. The doublet of doublets is split into a doublet of doublets of doublets by H(a)

1. H(c) is split into a doublet by H(d)

2. The doublet is split into a doublet of doublets by H(b)

- H(d) will also produce a doublet of doublet of doublets as shown below:

3. The doublet of doublets is split into a doublet of doublets of doublets by H(a)

1. H(d) is split into a doublet by H(c)

2. The doublet is split into a doublet of doublets by H(b)

Candidate III

We could draw a horizontal line of symmetry across the centre of the aromatic ring and predict that there would be two pairs of doublets as shown below where the line of symmetry is demonstrated by the dashed line:

H(a) would be split into a doublet by H(b) whilst H(b) would be split into a doublet by H(a). H(a') and H(b') would behave similarly and would appear at exactly the same chemical shifts as H(a) and H(b) respectively and each doublet would be if integral two.

This, however, would be incorrect as it takes no notice of the existence of the hydroxyl group and the withdrawing effects of the electronegative oxygen atom and the donation of the oxygen lone pair *into* the ring.

There is also the effect of the ester group which is electron donating to the ring. It is not affected by the presence of the oxygen atoms as there is a carbon atom between them and the oxygen atoms so the electronegativity of those oxygen atoms does not have an impact. The substituents distort the electronic structure of the ring and this means that there are should be two doublets of doublets of doublets both in the aromatic region, close together but not overlapping as shown below, firstly for H(a) and then for H(b) on the next page:

※ For H(a)

1. H(a) is split into a doublet by H(b)

3. The doublet of doublets is split into a doublet of doublets of doublets by H(a')

2. The doublet is split into a doublet of doublets by H(b')

The same applies to H(a').

■ For H(b) and thus also for H(b') we have the following situation:

1. H(b) is split into a doublet by H(a)

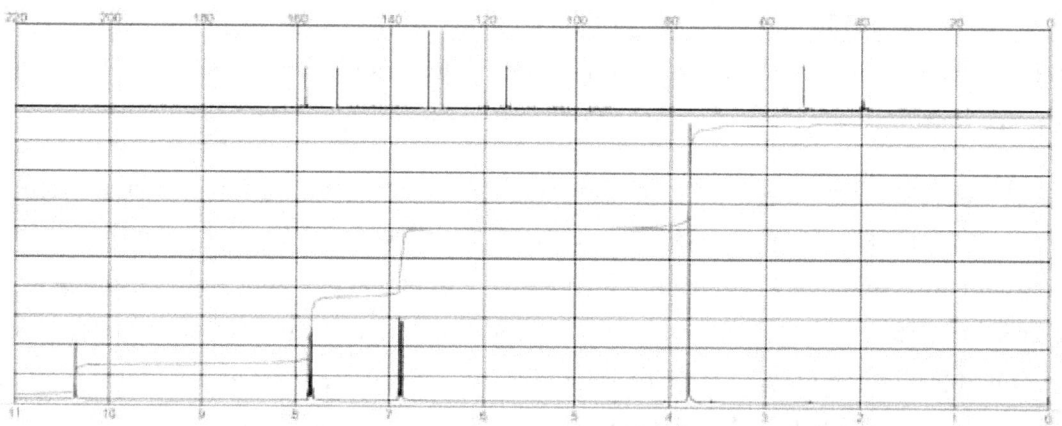

3. The doublet of doublets is split into a doublet of doublets of doublets by H(a')

2. The doublet is split into a doublet of doublets by H(b')

The significant issue here is that there will be only two doublets of doublets of doublets and each of the two multiplets will be of integral two. This is significantly different to Candidates I and II which will produce four doublets of doublets of doublets each of integral one.

If the observed spectrum contains four multiplets then we can discard Candidate III although distinguishing between Candidate I and II will require use of the coupling constants and the ^{13}C nmr spectrum but if the aromatic region of the ^{1}H nmr spectrum contains only two multiplets then it will be clear that this molecule, Candidate III, is the only possible structure.

The observed ^{1}H and ^{13}C nmr spectra are shown below.

There are peaks due to the – OH and – OCH_3 groups which would appear with all three candidates. It is perfectly clear, however, that, due to the existence of only two multiplets in the aromatic region (δ 6.2 – 8 ppm) the only possible structure is Candidate III.

We can confirm this by expanding the aromatic region

This expanded portion is shown on the next page.

The expanded ^{1}H NMR spectrum is shown below and it is perfectly clear that the two multiplets, each of integral two are doublets of doublets of doublets.

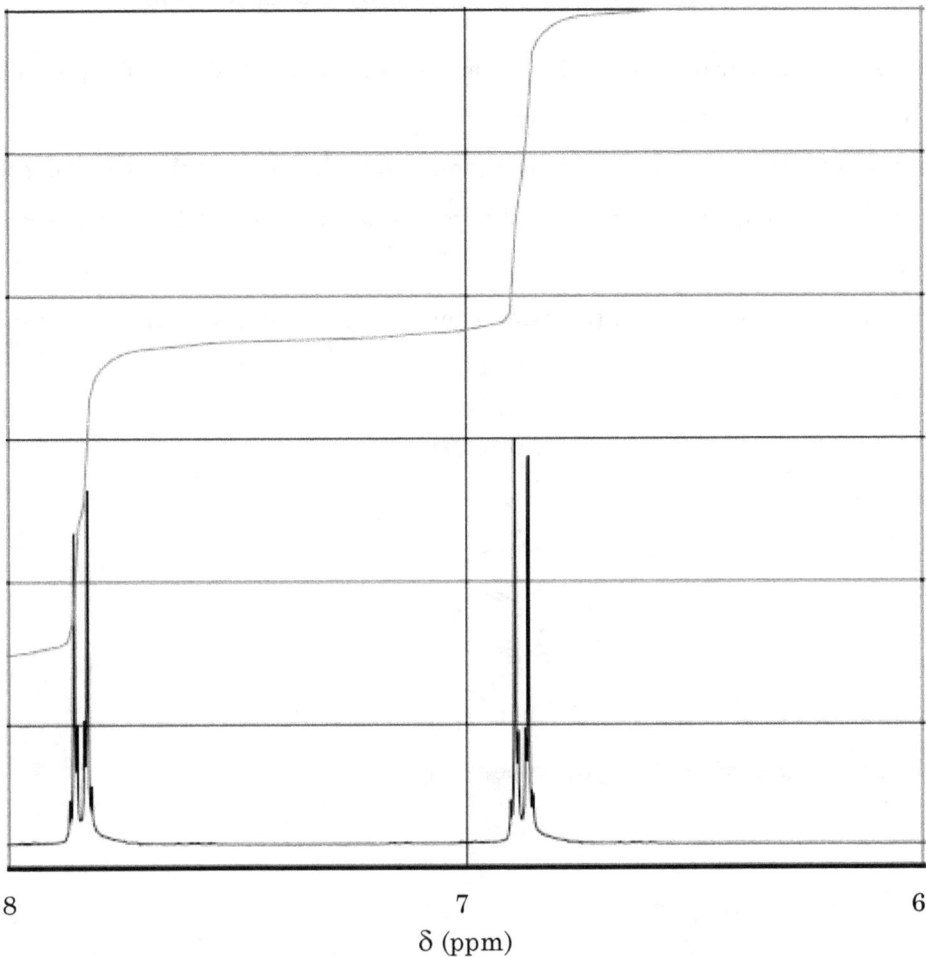

δ (ppm)

This spectrum can only be explained if the structure of the molecule is:

The coupling constants comprise J = 8, 3, and 0.5 Hz which is consistent with the prediction as these figures indicate the existence of hydrogen atoms in all of ortho − (J = 8 Hz), meta − (J = 3 Hz) and para- (J = 0.5 Hz) constituents. For example, using the labelling above, H(a) is ortho − to H(b), meta − to H(a') and para − to H(b).

The last task is to examine the ^{13}C nmr spectrum.

^{13}C NMR Spectrum

If the prediction is correct then we should expect to observe the following:

- Two peaks of integral two in the aromatic regions (δ 110 – 160 ppm) due to the unsubstituted carbon atoms in the ring;
- Two peaks of integral one also in that region assignable to the substituted carbon atoms;
- There will be a peak, of integral one, due to the methoxy (O – CH$_3$) in the region (δ 50 – 90 ppm) and
- One due to the carbon atom of the ester group which is bonded to the aromatic ring.

All of this is observed.

Conclusions

Structure:

Systematic Name: methyl 4-hydroxybenzoate.

Chapter III

Phenylethanol

Phenylethanol is found very widely in nature apparently because it is easily synthesised from the amino acid, phenylalanine.

It is found in cocoa beans, buckwheat, cooked mushrooms and many different species of flowers and acts as a pheromone for many species of butterflies and the cabbage looper moth (pictured above). With a honey-like and rose-type sweet odour it is also often added to floral fragrances.

With melting and boiling points of 21°C and 205°C, respectively, phenylethanol is, at room temperature, a colourless liquid which has slight solubility in water (1.95g dm^{-3}). It cools readily to form a colourless, crystalline solid.

This compound has the **elemental composition**: C: 78.62%, H: 8.27%, O: 13.10% and has the **formula mass (M_r)** of 122.1 g mol^{-1}.

This means that the empirical and molecular formula are both $C_8H_{10}O$.

Phenylethanol

Infrared Spectrum

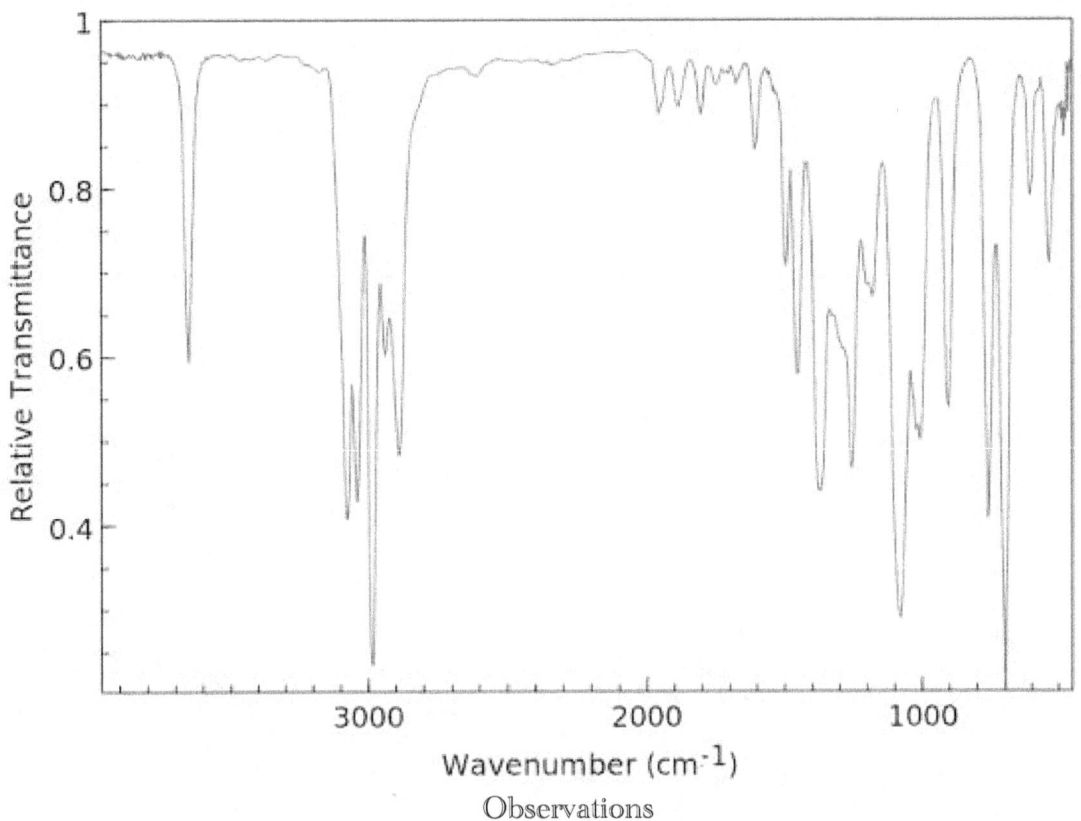

Observations

(√ / X)	Wavenumber range (cm⁻¹)	Wavenumber (cm⁻¹)	Assignment
√	3200 - 3700	3600	O – H
X	3200 - 3600		N – H
√	3000 – 3300	3080, 3020	C – H (aromatic)
√	2500 – 3000	2995, 2915, 2820	C – H (aliphatic)
X	2200 – 2500		C ≡ N
X	1700 – 1800		C = O
X	1600 – 1700		C = C (aliphatic)
X	1585 – 1600		C – C (aromatic)
X	1450 – 1600		C – C (aromatic)
√	1000 – 1300	1080	C – O
X	700 – 1000		C – X (X = Cl, Br or I)

Conclusions

This molecule must be aromatic but with an aliphatic substituent and a C – O bond. The sharp peak at 3600 cm⁻¹ is characteristic of a hydroxy compound and not an aliphatic alcohol. It cannot be an ester due to the absence of a C = O bond.

Phenylethanol

Mass Spectrum

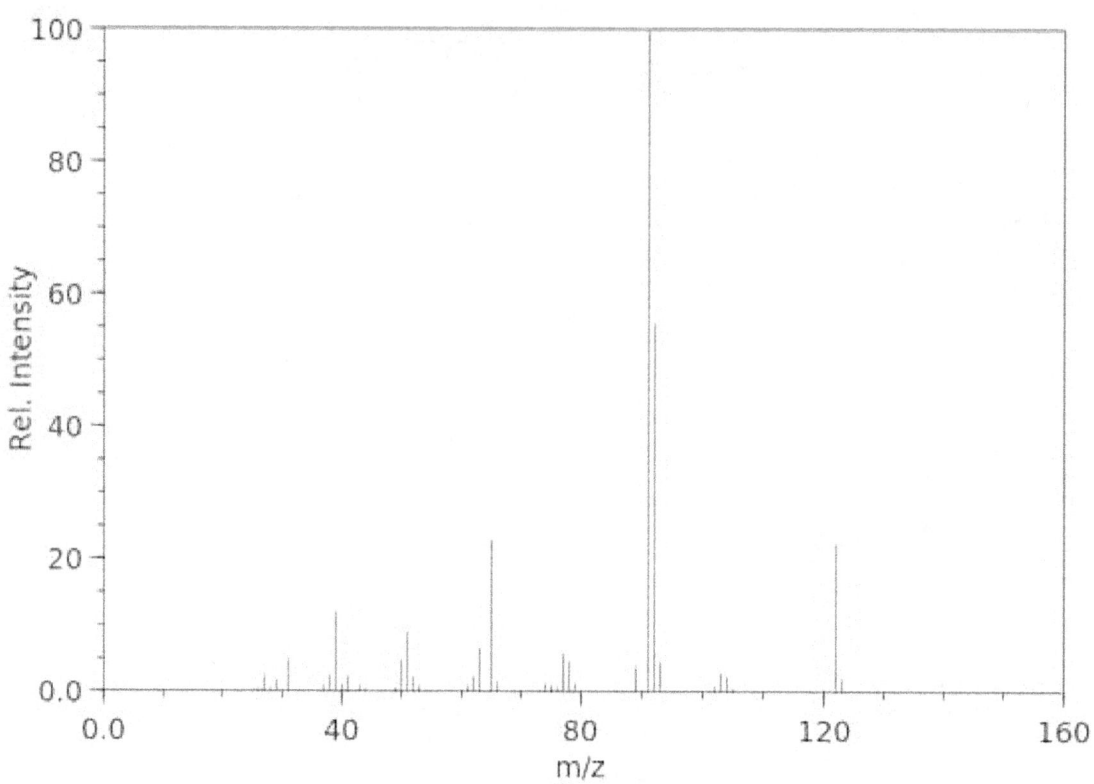

Observations

Charged fragments (m/z)	Assignment	Charged fragments (m/z)	Assignment
Molecular ion: 122	$[C_7H_{11}O]^+$	Base peak: 92	$[C_7H_8]^+$

77	$[C_6H_5]^+$	39	$[C_3H_3]^+$
65	$[C_5H_5]^+$	31	$[CH_3O]^+$
51	$[C_4H_3]^+$	29	$[C_2H_5]^+$ or $[CHO]^+$

Conclusions

The mass spectrum suggests the following:-

- The presence of the peak at m/z = 77 indicates the presence of a benzene ring with a single substituent;

- There is a functional group containing a –CH_3O group (m/z = 31) ;

Since there is only one substituent this means there must be a collection of atoms of formula C_2H_5O.

Phenylethanol

NMR Spectra

The peak at m/z=77 is conclusive evidence for the presence of a mono-substituted benzene ring and so there are two ways to assemble a benzene ring with a single functional group, – C_2H_5O is as follows:

I

CH_3
CH_2
O

II

HO–$\overset{H}{\underset{C}{}}$–$CH_3$

Although the structures of both candidates contain both a benzene ring and a C–O bond, the structure of Candidate I does not reconcile with the infrared or mass spectra since:-

- The infrared spectrum indicates the presence of an – OH functional group;
- There is no evidence from the mass spectrum of the existence of an – OCH_2CH_3 group.

This leaves us with only Candidate II as a plausible structure and we can predict the ^1H and ^{13}C nmr spectra before examining the actual spectrum.

^1H NMR Spectrum

There will be peaks assignable to the five hydrogen atoms attached to the ring which we consider next but we immediately predict the existence of:-

- A peak due to the hydroxyl (–OH) hydrogen atom and which can appear anywhere (δ 0 – 12 ppm). This is H(a) in the structure below and will be a doublet due to H(b).
- A doublet of quartet of integral one assignable to the $C_6H_5C(H)$– hydrogen atom, H(b) below which is split into a quartet by H(c) and the quartet is split into a doublet by H(a).
- There will be a signal, of integral three, due to the methyl (–CH_3) group, which, following the *n+1 rule*, will be a doublet due to the presence of a single hydrogen atom on the adjacent carbon atom, H(c).

^{a} HO–$\overset{b}{\underset{C}{H}}$–$CH_3$ ^{c}

We can now consider the signals due to the hydrogen atoms on the benzene ring.

21

Continuing the labelling used above, we can see that this is a fascinating structure. Working anti-clockwise from the left:-

■ H(d) is:-

   ■ Split into a doublet by H(e);

   ■ Which is split into a doublet of doublets by H(f);

   ■ Which is split into a doublet of doublets of doublets by H(e');

   ■ Which is split into a doublet of doublets of doublets of doublets by H(f).

as shown below

1. H(d) is split into a doublet by H(e)

2. This doublet is split into a doublet of doublets by H(f)

4. This doublet of doublets of doublets is split into a doublet of doublets of doublets of doublets by H(d')

3. This doublet of doublets is split into a doublet of doublets of doublets by H(e')

Exactly the same conclusion would be drawn if we worked clockwise from H(d') and this means we should predict a doublet of doublets of doublets of doublets of integral two assignable to H(d) / H(d').

■ H(e):

   ■ Is split into a doublet by H(d);

   ■ This doublet is split into a doublet of doublets by H(f) which is;

   ■ Split into a doublet of doublets of doublets by H(e') which is;

   ■ Split into a doublet of doublets of doublets of doublets by H(d')

as shown below:

1. H(e) is split into a doublet by H(d)

2. This doublet is split into a doublet of doublets by H(f)

4. This doublet of doublets of doublets is split into a doublet of doublets of doublets of doublets by H(d')

3. This doublet of doublets is split into a doublet of doublets of doublets by H(e')

22

* H(f), integral one,:-

    * Is split into a triplet by H(e) and H(e')

    * Which is split into a triplet of triplets by H(d) and H(d')

To summarise then we have the following predictions in which we can predict the region but not the exact chemical shift of the peaks.

Chemical shift δ (ppm)	Assignment(s)	Integral	Multiplicity
0 − 12	H(a)	1	Doublet
3 − 4.5	H(b)	1	Doublet of quartets
0.5 − 2	H(c)	3	Doublet
6.2 − 8	H(d) & H(d')	2	Doublet of doublet of doublets of doublets
6.2 − 8	H(e) & H(e')	2	Doublet of doublet of doublets of doublets
6.2 − 8	H(f)	1	Triplet of triplets

We also predict the coupling constants of the aromatic hydrogen atoms:-

* H(d) is ortho- to H(e), meta- to H(d') and H(f), and para- to H(e')
* H(e) is ortho- to H(d) and H(f), meta to H(e') and para- to H(d')
* H(f) is ortho- to H(e) and H(e') and meta- to H(d) and H(d')

We are now in a position to consider the actual, observed ^{1}H nmr spectrum and compare the predictions of chemical shift, integral, multiplicity and coupling constant to the prediction.

Phenylethanol

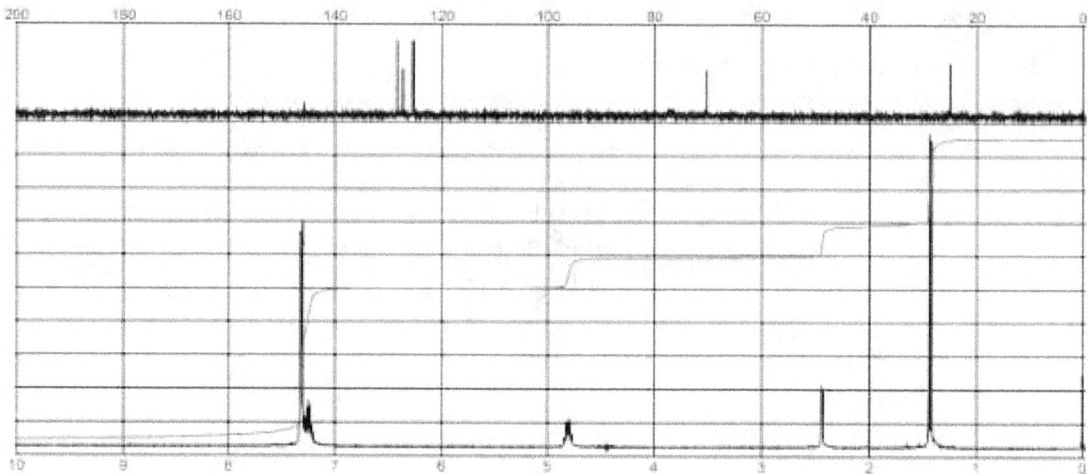

We can see immediately that there are four multiplets which is commensurate with the proposed structure as highlighted in the structure below using the labelling from before:

From the data sheet we can immediately conclude that:

- The doublet at δ 1.45 ppm, of integral three, can be assigned to H(c) as it split by H(b).
- The quartet at δ 4.85 ppm, of integral one, is produced by H(b) as it is split by H(c).
- The complex multiplet in the aromatic region, of total integral five, must due to the hydrogen atoms bonded to the ring, H(d), H(d'), H(e), H(e') and H(f) enclosed in the dashed box.

This means that the doublet at δ 2.45 ppm, of integral one, must be to H(a) which is split by H(b).

There is no other structure which is consistent with all the measurements made so far in the infrared, the fragmentations in the mass spectrum and, of course, the ^{1}H nmr spectrum. There is one task remaining which is to consider the ^{13}C nmr spectrum which we do next.

Phenylethanol

If we consider the molecule again, with new labelling and with reference to the data chart, we can see that it will contain the following carbon atoms:

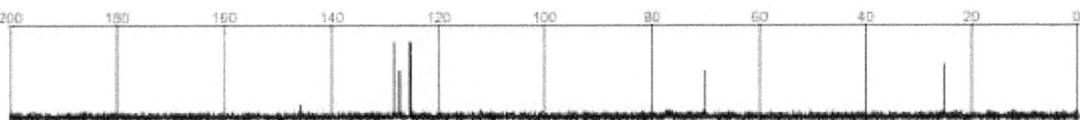

- C(g) will produce a singlet of integral one in the region δ 0 – 50 ppm.
- C(h) will produce a singlet of integral one in the region δ 50 – 90 ppm.
- There will be three singlets in the aromatic region, δ 110 – 160 ppm as follows:
  - One singlet of integral two due to C(i) and C(i') which are mutually chemically and magnetically equivalent.
  - One singlet of integral two assignable to C(j) and C(j') which are also mutually chemically and magnetically equivalent to each other.
  - One singlet, of integral one, which can be assigned to C(k).

If we reconsider the observed ¹³C nmr spectrum, this is exactly what is observed:

and we can tabulate the following assignments:

Chemical shift δ (ppm)	Assignment(s)	Integral
24	C(g)	1
50	C(h)	1
124	C(i) & C(i') or C(j) & C(j')	2
126	C(k)	1
129	C(i) & C(i') or C(j) & C(j')	2

which leads us to make the following conclusions:

Phenylethanol

Conclusions

Structure:

Systematic name: 1-phenylethan-1-ol.

Chapter IV

Syringaldehyde

Widely found in nature, **syringaldehyde** is used as a trail pheromone by ants.

It has also been suggested that the, female, European elm bark beetle (pictured above) uses it to find a suitable tree on which to lay its eggs.

A colourless solid with a melting point of 110°C and boiling point of 193°C, and almost insoluble in water, it has the **elemental composition**: C: 59.31%, H: 5.55%, O: 35.15% and a **formula mass (M_r)** of 182.10 g mol^{-1}.

These means that the empirical and molecular formula are both $C_9H_{10}O_4$.

The presence of more than six carbon atoms implies the molecule is a long chain alkane, a substituted shorter chain or a substituted benzene ring. The presence of four oxygen atoms also implies the possibility of the molecule being an ester or a carboxylic acid.

Syringaldehyde

Infrared Spectrum

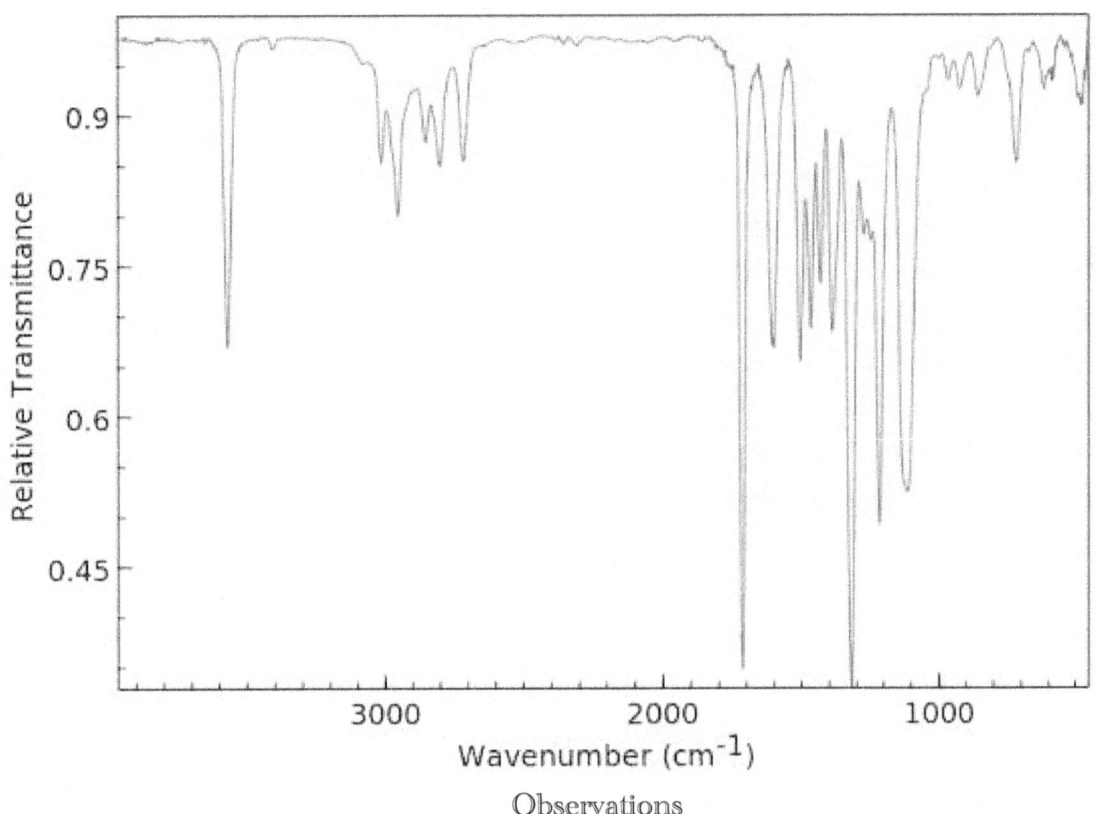

Observations

(√ / X)	Wavenumber range (cm⁻¹)	Wavenumber (cm⁻¹)	Assignment
√	3200 - 3700	3520	**O – H**
X	3200 - 3600		**N – H**
√	3000 – 3300	3040	**C – H (aromatic)**
√	2500 – 3000	Four peaks	**C – H (aliphatic)**
X	2200 – 2500		**C ≡ N**
√	1700 – 1800	1710	**C = O**
X	1600 – 1700		**C = C (aliphatic)**
X	1585 – 1600		**C – C (aromatic)**
X	1450 – 1600		**C – C (aromatic)**
√	1000 – 1300	1320, 1200 or 1100	**C – O**
X	700 – 1000		**C – X** (X = Cl, Br or I)

Conclusions

This molecule is aromatic with aliphatic groups as well as a carbonyl (C=O) and a C – O bond. This suggests that the molecule is a carboxylic acid. The peak at 3520 cm⁻¹ implies the presence of a phenolic functional group.

Syringaldehyde

Mass Spectrum

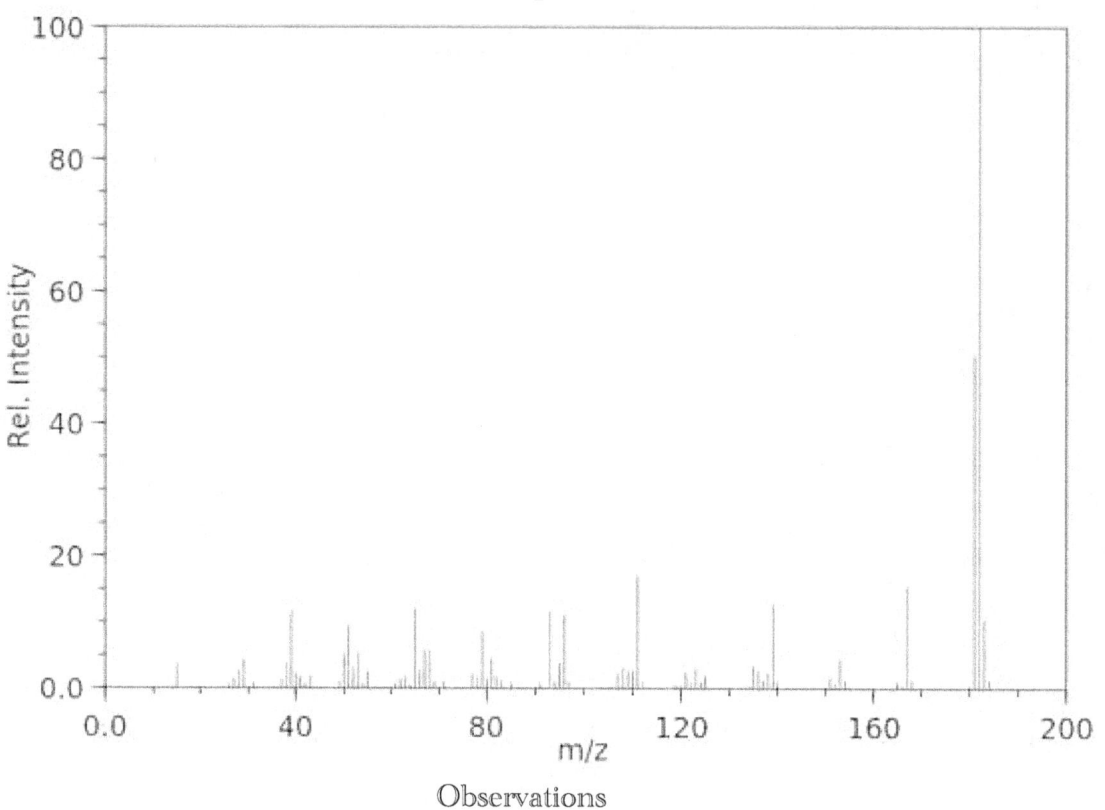

Observations

Charged fragments (m/z)	Assignment	Charged fragments (m/z)	Assignment
Molecular ion: 182	$[C_9H_{10}O_4]^+$	**Base peak: 182**	$[C_9H_{10}O_4]^+$
167	$[C_8H_7O_4]^+$	79	$[C_6H_5 + 2H]^+$
129	$[C_6H_7O_2\}^+$	65	$[C_5H_5]^+$
111	$[C_7H_{11}O_2]^+$	51	$[C_4H_9]^+$
86	$[C_5H_{10}O]^+$	39	$[C_3H_9]^+$
81	$[C_6H_9]^+$	29	$[C_2H_5]^+$ & $[CHO]^+$

Conclusions

▪ The mass spectrum is difficult to interpret but the peak at m/z = 29 implies the existence of a –CHO functional group.

▪ The infrared spectrum infers the existence of an aromatic ring and the peak at m/z = 79 supports this supposition.

We can learn far more from the nmr spectra which we next starting with the ^{1}H NMR spectrum.

NMR Spectra

There are a number of possible isomers of molecular formula $C_9H_{10}O_4$. There is a sensible tendency to consider that a molecule with two or more oxygen atoms could be a carboxylic acid or an ester. There are, however, many other possibilities include a compound with some hydroxy, $-OH$, groups as well as ethers. Another important consideration, however, is the number of carbon atoms in the molecular formula since anything above six suggest the possibility of a benzene ring. If there is a benzene ring then that accounts for six of the carbon atoms, leaving the remainder of the molecule, $-C_3H_{10}O_4$, to be fitted in. This suggests that there might be one or two ester groups attached and the list of possible structures continues.

If we return to the infrared spectrum, we have already noted that the molecule contains a :-

- Phenolic, $-OH$, group;
- Benzene ring;
- Carbonyl, $C=O$, group and;
- $C - O$ bond.

There is evidence for the presence of the phenolic group in the infrared region so we can make the base structure as shown below:

where the ∿ is a placeholder for the other groups.

Since there is no aliphatic carboxylic acid, $-COOH$, functional group (there is no broad and strong hydroxyl peak) then there must be at least some of the following:-

- aldehyde, $-CHO$, functional groups;
- ester $C-COOR$ groups (where R is an alkyl group); or
- alkoxy, $-OR$, groups

somewhere on the ring or a combination of the above.

Since there are a number of possible combinations it is best to examine the ^1H nmr spectrum for some clues. The observed spectrum is displayed on the next page.

Syringaldehyde

^{1}H NMR Spectrum

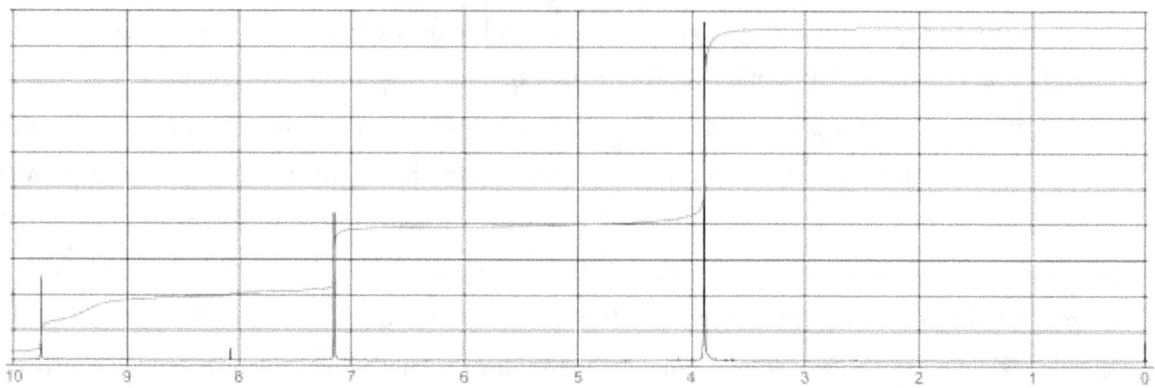

Whatever the structure of this molecule, this spectrum is astonishingly and intriguingly simple since it merely comprises four singlets including one that is very small. We can make the following observations and assignments:

Chemical shift δ (ppm)	Integral	Multiplicity	Assignment(s)
9.75	1	Singlet	–CHO
8.15	<1	Singlet	–OH ?
7.2	2	Singlet	Aromatic
3.88	6	Singlet	–OCH$_3$

The singlet of integral less than one is a signal we will come back to.

▪ The singlet at δ 7.2 ppm confirms the presence of a benzene ring

▪ That there are only two hydrogen atoms on the ring means that the basic structure must be of general form:

where R, R', R" and R'" are any functional group but cannot be a hydrogen atom

The aromatic signal is a singlet and so there is no difference between the two hydrogen atoms. This means that only the left hand isomer is a feasible structure.

The next task is to identify the nature of the substituents R, R', R" and R'".

■ From the ^{1}H nmr spectrum we know that one of the substituents is the –CHO group which presents a signal at δ 9.75 ppm so we now have two possible structures:

We also know that there are six hydrogen atoms of the general form, –OCH$_3$. They cannot form a group of structure –OCH$_2$H$_3$ as the number of hydrogen does not fit and there is no multiplicity in the signal.

This brings us a further development of the structure to these two candidates:

■ We also have one more hydrogen and one more oxygen atom to place.

These must form an –OH functional group for two reasons:

   ■ There is only spare place on each of the candidate molecules;

   ■ The infrared spectrum exhibits a peak at 3520 cm^{-1} which is characteristic only of an aromatic phenolic group. Phenolic hydrogen atoms exchange very rapidly in solution and this explains why the peak due to the –OH hydrogen atom appears with an integral below one.

This produces the two final candidates:

Our final task is to determine which one is correct.

To distinguish between the two candidates we must return to the mass spectrum.

Syringaldehyde

If we imagine the molecule fragmenting we imagine the following fractures where the fragments are marked by the dashed lines with the m/z values for each fragment stated in each section:

m/z = 97   CHO
    H   H
    H₃CO   OCH₃
    OH   m/z = 85

m/z = 67   CHO
    H   H
    H₃CO   OCH₃
    OH   m/z = 115

All four of these fragments appear in the mass spectrum. We can, thus, be confident that this is the correct structure but our final task is to predict and examine the ¹³C nmr spectrum.

¹³C NMR Spectrum

If this structure is correct then we can expect to observe the following:-

Chemical shift δ (ppm)	Integral	Assignment(s)
160 – 220	1	Ar – CHO
110 – 160	2	**Ar – H**
110 – 160	2	**Ar – OCH₃**
110 – 160	1	**Ar - CHO**
110 – 160	1	**Ar – OH**
50 – 90	2	Ar – OCH3

where Ar represents a benzene ring and **Ar** represents a carbon atom **in** the ring.

All of this is observed as shown below:

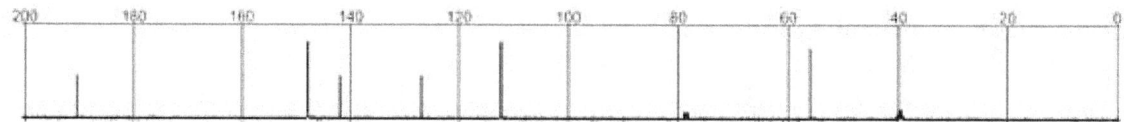

Overall Conclusions

Structure:

    CHO
   H   H
  H₃CO   OCH₃
    OH

Systematic name: 4-hydroxy-3,5-dimethoxybenzaldehyde

Chapter V

Benzyl acetate

Benzyl acetate has a sweet aroma and taste and is a constituent of jasmine, apples and pears. It confers a jasmine or apple type fragrance to many personal hygiene products and is one of many compounds which act as an attractant to male orchid bees (pictured above). This has led it to be used as bait to collect bees.

It is a colourless liquid with a melting point of -51°C and a boiling point of 212°C and is almost completely insoluble in water.

This compound has the **elemental composition**: C: 71.95%, H: 6.70%; O: 21.32% and has a **formula mass (M_r)** of 150.14 g mol⁻¹.

This means that the empirical formula and molecular formulas are both $C_9H_{10}O_2$.

The number of carbon atoms implies a long, or a shorter substituted, aliphatic chain or an aromatic compound whilst the presence of two oxygen atoms implies that the molecule is a carboxylic acid or an ester. The compound has a sweet aroma which is characteristic of esters and so the assumption that it is an ester should be our starting point.

Benzyl acetate

Infrared Spectrum

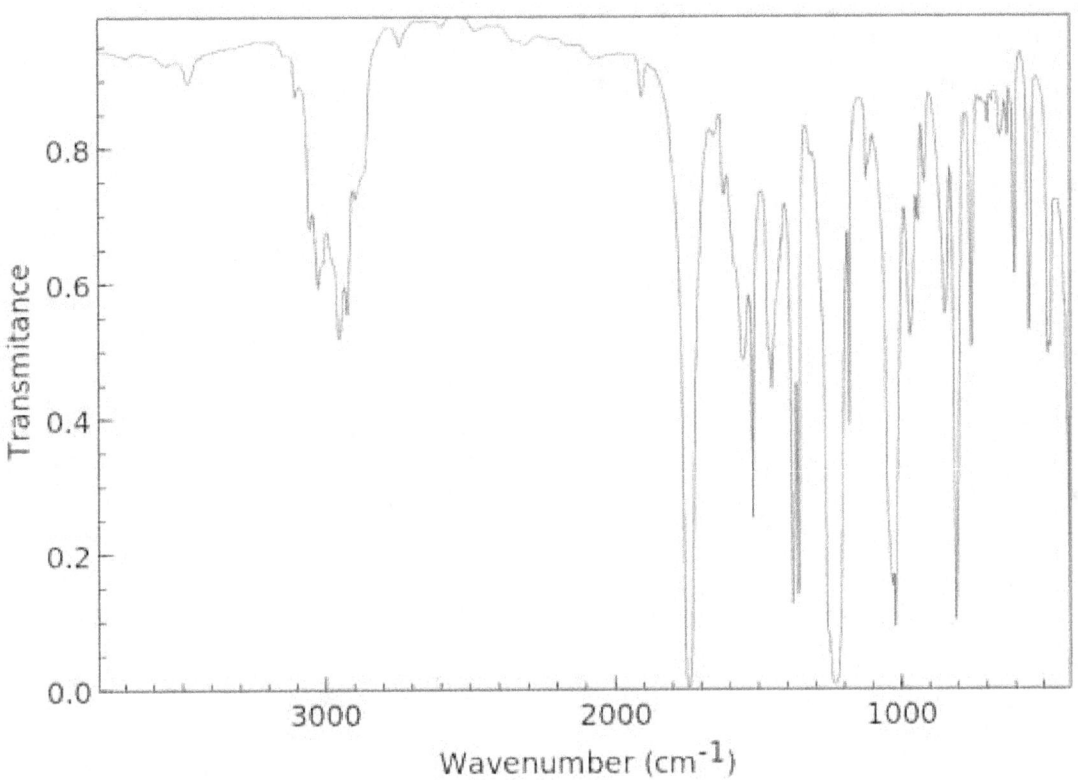

Observations

($\sqrt{}$ / X)	Wavenumber range (cm^{-1})	Wavenumber (cm^{-1})	Assignment
X	3200 - 3700		O – H
X	3200 - 3600		N – H
$\sqrt{}$	3000 – 3300	3020, 3045	C – H (aromatic)
$\sqrt{}$	2500 – 3000	2940,2880	C – H (aliphatic)
X	2200 – 2500		C $\equiv$ N
$\sqrt{}$	1700 – 1800	1740	C = O
X	1600 – 1700		C = C (aliphatic)
X	1585 – 1600		C – C (aromatic)
X	1450 – 1600		C – C (aromatic)
$\sqrt{}$	1000 – 1300	1230	C – O
X	700 – 1000		C – X (X = Cl, Br or I)

Conclusions

This molecule contains an aromatic ring, an aliphatic group and, due to the C = O and the C – O group but, with no hydroxyl group, it must also contain an ester functional group.

Benzyl acetate

Mass Spectrum

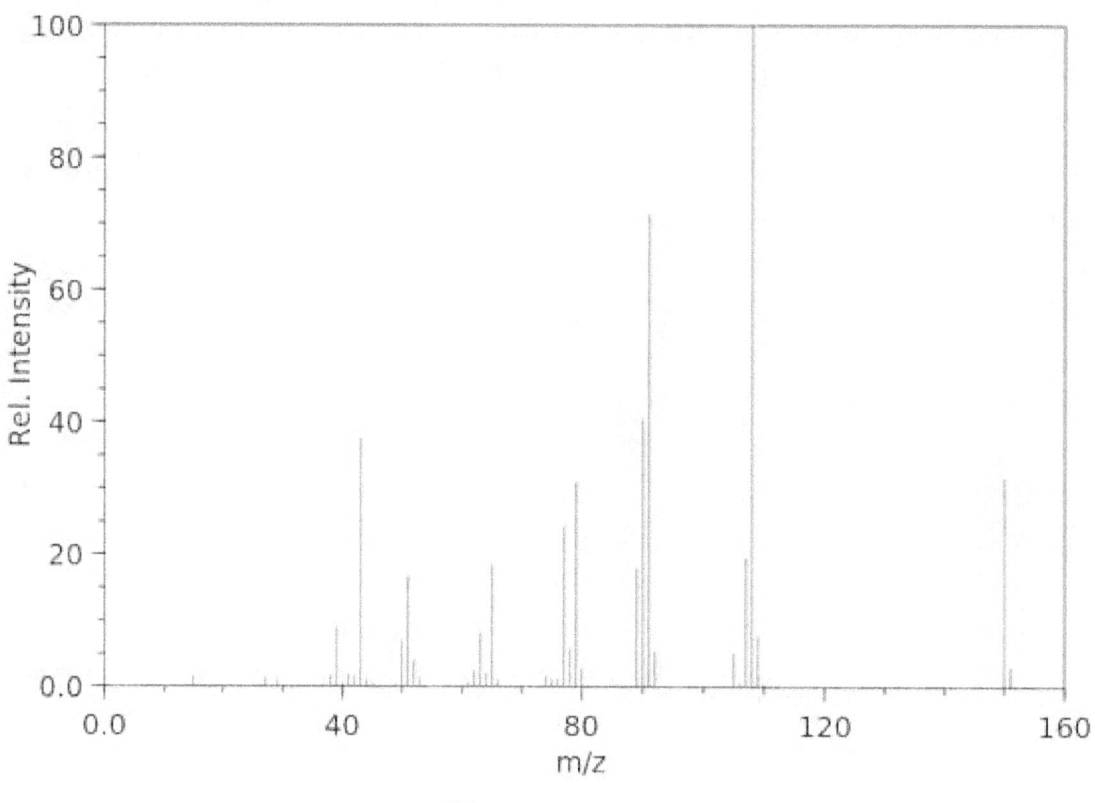

Observations

Charged fragments (m/z)	Assignment	Charged fragments (m/z)	Assignment
Molecular ion: 150	$[C_9H_{10}O_2]^+$	Base peak: 108	$[C_7H_8O]^+$
91	$[C_7H_7]^+$	51	$[C_4H_3]^+$
77	$[C_6H_5]^+$	43	$[C_3H_7]^+$
65	$[C_5H_5]^+$	29	$[C_2H_5]^+$

Conclusions

The mass spectrum adds little to our understanding but suggests that there is a benzene ring in the molecule due to the characteristic m/z =77 peak which is only ever caused by a mono-substituted benzene ring. We also know that there is a C = O and a C − O bond suggesting that the ring has an ester group attached to it. That group cannot be a carboxylic acid functional group as the infrared spectrum demonstrates the *absence* of a C − O − H group.

We can determine the structure from the ^{1}H and ^{13}C nmr spectra which we consider next.

NMR Spectra

The nmr spectra are fascinating and relatively simple to analyse.

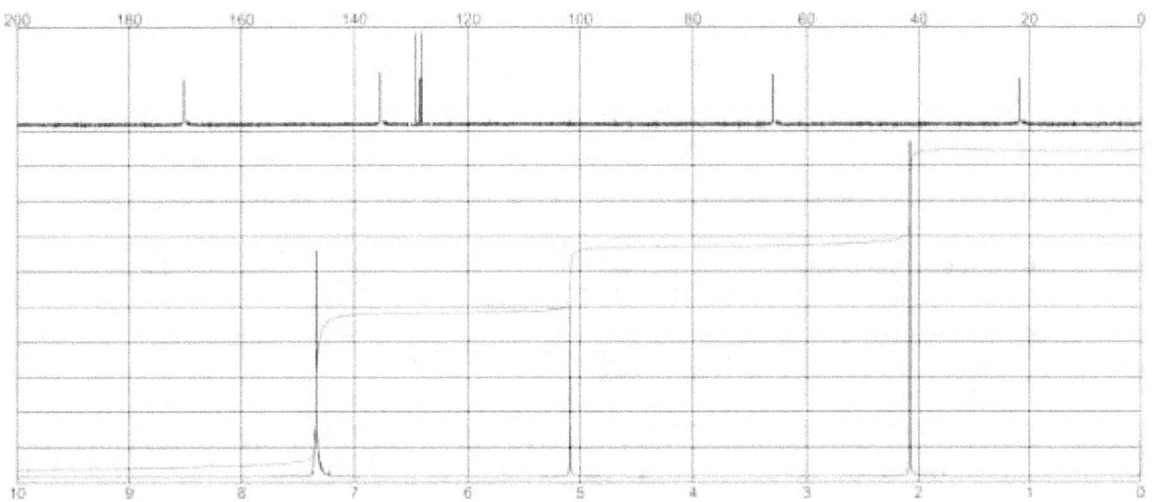

¹H NMR Spectrum

We can make the following observations from the ¹H nmr spectrum as follows:

Chemical shift δ (ppm)	Assignment(s)	Integral	Multiplicity
7.4	Aromatic hydrogen atoms	5	Complex multiplet
5.1	$CH_2 - O$	2	Singlet
2.1	$- CH_3$	3	Singlet

Since there is one multiplet, of integral five, this indicates that there can only be one substituent on the aromatic ring so it is of general form, where R is the remainder of the molecule:

The two other signals indicate the presence of an ester grouping, which concurs with the infrared and mass spectra so the only possible structure is:

We can confirm or disprove this structure by predicting the ¹³C nmr spectrum and examining it and we do this next.

^{13}C NMR Spectrum

If the proposed structure

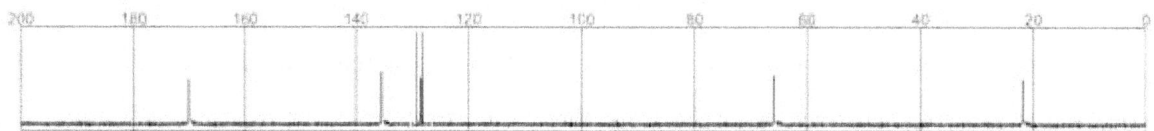

is correct then we can expect to observe the following signals in the ^{13}C nmr spectrum:

Chemical shift δ (ppm)	Assignment(s)	Integral
160 – 220	C = O	1
115 – 160	Aromatic carbons	6
50 – 90	C – O	1
0 – 50	– CH$_3$	1

This is exactly what we observe in the spectrum which is repeated below:

so the structure is confirmed.

Conclusions

Structure:

Systematic name: Benzyl ethanoate

Chapter VI

Creosol

Creosol is used as a flavouring and is responsible for the natural flavouring of buckwheat. It has an odour somewhat similar to phenol and creosote, of which it is a major constituent, and is employed as a foraging pheromone by beetles (pictured above), houseflies and millipedes (pictured below).

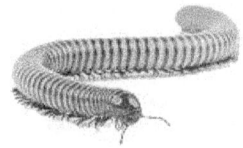

This colourless compound has a melting point of 6°C and a boiling point of 221°C and so, at room temperature, it is a colourless liquid.

It has the following other properties:

Elemental composition: C: 69.48%, H: 7.31%; O: 23.16%

Formula mass (M_r): 138.16 g mol^{-1}.

These means that the:

Empirical formula is C_4H_5O

Molecular formula is $C_8H_{10}O_2$.

As with some other pheromones in this volume, the presence of more than six carbon atoms suggests either an aliphatic chain or an aromatic ring whilst the presence of two oxygen atoms immediately suggests that it is a carboxylic acid or an ester.

Creosol

Infrared Spectrum

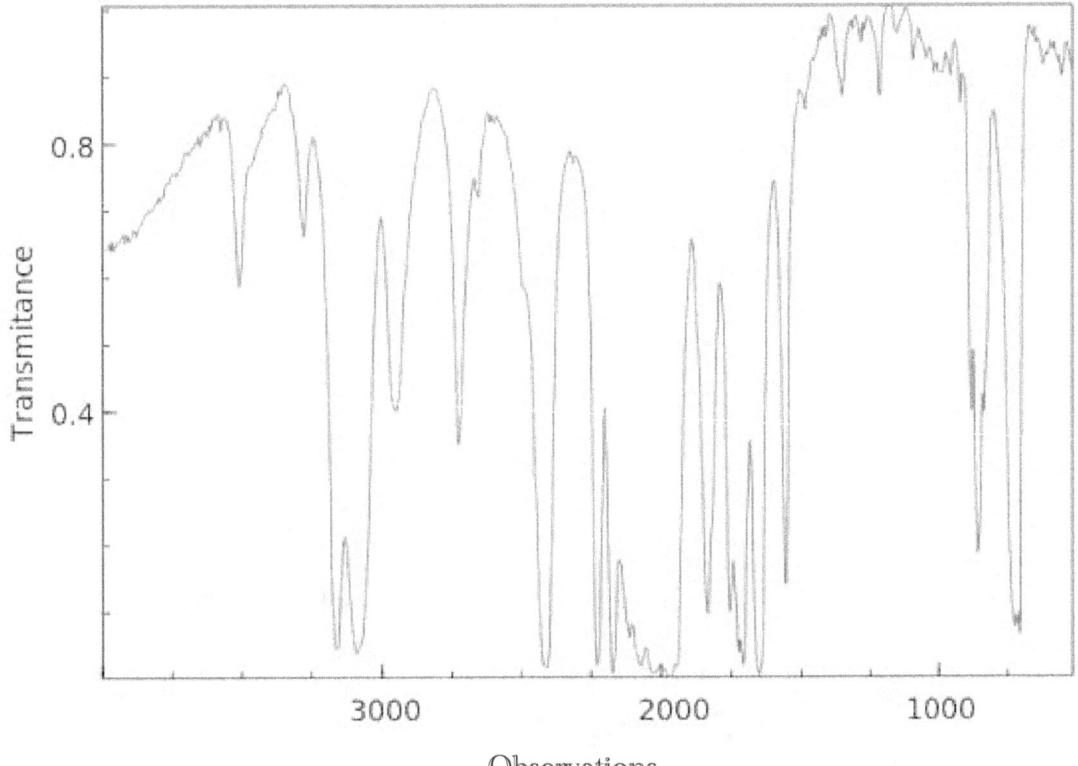

Observations

(√ / X)	Wavenumber range (cm⁻¹)	Wavenumber (cm⁻¹)	Assignment
√	3200 - 3700	3480	O – H
X	3200 - 3600		N – H
√	3000 – 3300	3150, 3075	C – H (aromatic)
√	2500 – 3000	2930	C – H (aliphatic)
X	2200 – 2500		C ≡ N
X	1700 – 1800		C = O
X	1600 – 1700		C = C (aliphatic)
√	1585 – 1600	1590	C – C (aromatic)
√	1450 – 1600	1520	C – C (aromatic)
√	1000 – 1300	1250	C – O
X	700 – 1000		C – X (X = Cl, Br or I)

Conclusions

This compound is aromatic and has an – OH group attached as well as alkyl substituents. The presence of a C – O bond but the absence of a C = O peak indicates that the molecule cannot be an ester or a carboxylic acid and there must be an alkoxy group substituent.

Creosol

Mass Spectrum

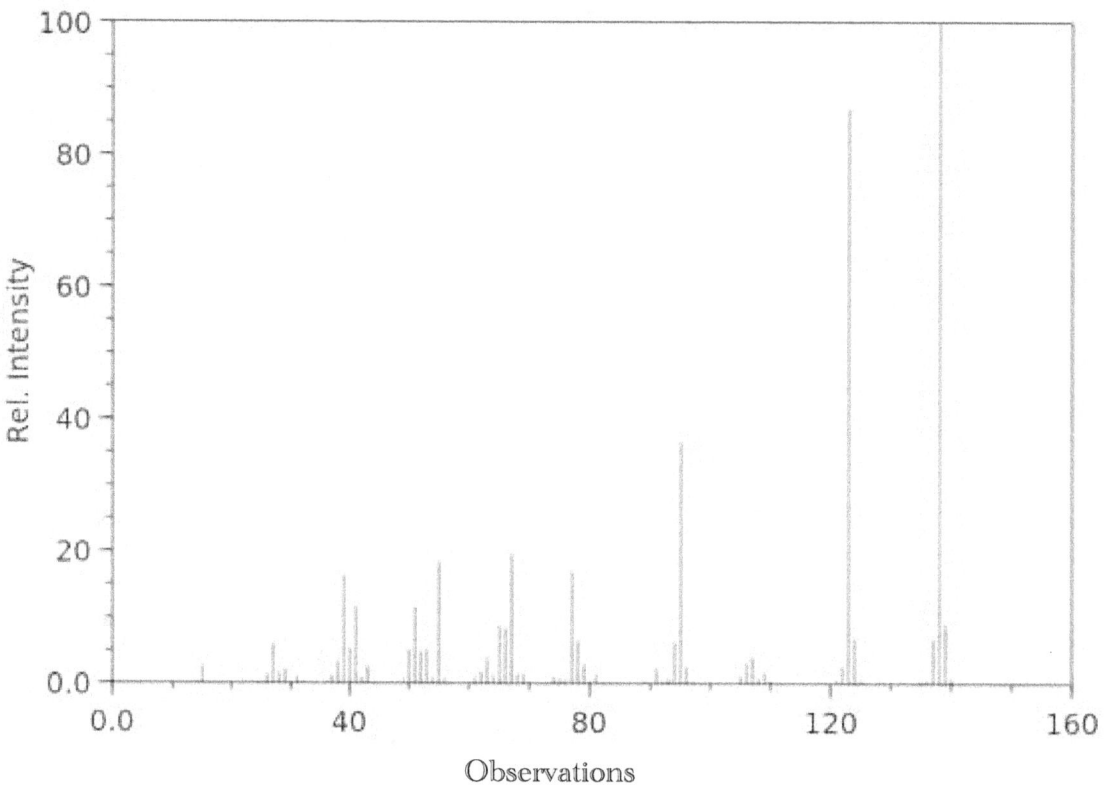

Observations

Charged fragments (m/z)	Assignment	Charged fragments (m/z)	Assignment
Molecular ion: 138	$[C_8H_{10}O_2]^+$	**Base peak: 138**	$[C_8H_{10}O_2]^+$

123	$[C_8H_{10}O]^+$	67	$[C_5H_7]^+$
108	$[C_7H_8O]^+$	55	$[C_4H_7]^+$
95	$[C_6H_7O]^+$	39	$[C_3H_3]^+$

Conclusions

This spectrum indicates that the molecule must contain a benzene ring since, although there are sufficient carbon atoms to form a long chain, there are insufficient hydrogen atoms to form a molecular formula. If we accept that there is a benzene ring then there must, from the infrared spectrum also be an –OH group, a methyl group and an – O – CH_3 group attached to the ring.

There are a number of possible isomers of this molecule which are discussed next when we consider the ^{1}H and ^{13}C nmr spectra.

NMR Spectra

Since this molecule contains a benzene ring, an –OH group, a methyl (– CH_3) group and an – O – CH_3 group attached to the ring then we have the following candidate isomers:

The two most right hand images are mirror images of each and can be interconverted simply by flipping one over and so we have seven candidates.

The only way to distinguish between the candidate molecules will be to examine the aromatic region of the ^1H nmr spectrum since the coupling constants indicate the relative (ortho –, meta –, para) positions of the aromatic hydrogen atoms.

We will examine each candidate in turn with the hydrogen atoms labelled alphabetically.

Candidate I

- H(a) is ortho – to H(b) and meta – to H(c),
- H(b) is ortho – to both H(a) and H(c) and so there will be coupling constants in both of the ranges J = 7 – 10 Hz (ortho–) and 2 – 3 Hz (meta –).
- H(c) is ortho – to H(b) and meta – to H(a).

Multiplicities:

- Both H(a) and H(c) will produce a doublet of doublets with two coupling constants, one in the range J = 7 – 10 Hz and one in the range J = 2 – 3 Hz.
- H(b) will also produce a doublet of doublets and the coupling constants will both be in the region J = 7 – 10 Hz.

Candidate II

H(a) is ortho – to H(b) and H(a) is para – to H(d) whilst H(b) is meta – to H(d) .

This means that there will be coupling constants in all of the ranges J = 7 – 10 Hz (ortho–), J = 2 – 3 Hz (meta –) and J = 0 – 2 Hz (para –).

Multiplicities:

▪ H(a) will produce a doublet of doublets with two coupling constants, with one in the range J = 7 – 10 Hz (ortho–) and one in the range J = 0 – 2 Hz (para–).

▪ H(b) will produce a doublet of doublets with two coupling constants with one in the range J = 7 – 10 Hz (ortho–) and one in the range J = 2 – 3 Hz (meta–).

▪ H(d) will produce a doublet of doublets with two coupling constants with one in the range J = 2 – 3 Hz (meta–) and one in the range J = 0 – 2 Hz (para–).

Candidate III

H(a) is meta – to H(c), H(c) is ortho – to H(d) and H(a) is para – to H(d).

This means that there will be coupling constants in all of the ranges J = 7 – 10 Hz (ortho–), J = 2 – 3 Hz (meta –) and J = 0 – 2 Hz (para –).

Multiplicities:

▪ H(a) will produce a doublet of doublets with two coupling constants, with one in the range J = 7 – 10 Hz (meta–) and one in the range J = 0 – 2 Hz (para–).

▪ H(c) will produce a doublet of doublets with two coupling constants with one in the range J = 7 – 10 Hz (ortho–) and one in the range J = 2 – 3 Hz (meta–).

▪ H(d) will produce a doublet of doublets with two coupling constants with one in the range J = 7 – 10 Hz (meta–) and one in the range J = 0 – 2 Hz (para–).

Candidate IV

H(a) is ortho – to H(b) and meta – to H(c) whilst H(b) is ortho – to H(c) and H(a) and H(c) is ortho – to H(b) and meta – to H(a). This means that there will be coupling constants in both of the ranges J = 7 – 10 Hz (ortho–), J = 2 – 3 Hz (meta –) but no coupling constants in the range J = 0 – 2 Hz, due to the absence of para– substituents.

Multiplicities:

* H(a) will produce a doublet of doublets with two coupling constants, with one in the range J = 7 – 10 Hz (ortho–) and one in the range J = 2 – 3 Hz (meta–).

* H(b) will produce a doublet of doublets with two coupling constants but both in the same range, J = 7 – 10 Hz, as H(b) is ortho – to both H(a) and H(c) with one in the range J = 7 – 10 Hz (ortho–) and one in the range J = 2 – 3 Hz (meta–).

* H(c) will produce a doublet of doublets with two coupling constants with one in the range J = 7 – 10 Hz (ortho–) and one in the range J = 2 – 3 Hz (meta–).

Candidate V

H(a) is ortho – to H(b) and para – to H(d) , H(b) is meta – to H(c) and H(p) is meta – to H(c). This means that there will be coupling constants in all of the ranges J = 7 – 10 Hz (ortho–), J = 2 – 3 Hz (meta –) and J = 0 – 2 Hz (para –).

Multiplicities:

* H(a) will produce a doublet of doublets with two coupling constants, with one in the range J = 7 – 10 Hz (ortho–) and one in the range J = 0 – 2 Hz (para–).

* H(b) will produce a doublet of doublets with one coupling constant in the range, J = 7 – 10 Hz and one in the range J = 2 – 3 Hz.

* H(d) will produce a doublet of doublets with two coupling constants with one in the range J = 2 – 3 Hz (ortho–) and one in the range J = 0 – 2 Hz (para–).

Candidate VI

H(a) is meta– to H(c) and para– to H(d) whilst H(c) is ortho – to H(d).

This means that there will be coupling constants in all of the ranges J = 7 – 10 Hz (ortho–), J = 2 – 3 Hz (meta–) but none in the range J = 0 – 2 Hz (para–).

Multiplicities:

▪ H(a) will produce a doublet of doublets with two coupling constants, with one in the range J = 2 – 3 Hz (meta–) and one in the range J = 0 – 2 Hz (para–).

▪ H(c) will produce a doublet of doublets with one coupling constant in the range, J = 2 – 3 Hz and one in the range J = 7 – 10 Hz.

▪ H(d) will produce a doublet of doublets with two coupling constants with one in the range J = 7 – 10 Hz (ortho–) and one in the range J = 0 – 2 Hz (para–).

Candidate VII

H(b) is ortho– to H(c) and meta– to H(d) whilst H(c) is ortho – to both H(b) and H(d). This means that there will be two coupling constants, J = 7 – 10 Hz (ortho–), and J = 2 – 3 Hz (meta–) but none in the range J = 0 – 2 Hz (para–).

Multiplicities:

▪ H(b) will produce a doublet of doublets with two coupling constants, with one in the range J = 7 – 10 Hz (ortho–) and one in the range J = 2 – 3 Hz (meta–).

▪ H(c) will produce a doublet of doublets with one coupling constant in the range, J = 2 – 3 Hz as it is meta– to both H(b) and H(d).

▪ H(d) will produce a doublet of doublets with two coupling constants with one in the range J = 7 – 10 Hz (ortho–) and one in the range J = 2 – 3 Hz (meta–).

Creosol

¹H NMR Spectrum

It is now time to examine the actual spectrum.

The ¹H nmr spectrum of all of these molecules will contain the following:-

- A singlet, of integral one, due to the – OH group which can appear anywhere in the spectrum;
- A singlet, of integral three, in the region δ 3 – 4.2 ppm due to the – OCH_3 group;
- A singlet of integral three, in the region δ 0.5 – 2 ppm due to the – CH_3 group;
- One or more multiplets, of total integral three, in the δ 6 – 8 ppm region due to the three hydrogen atoms bonded directly to the aromatic ring.

The ¹H nmr spectrum is shown below:

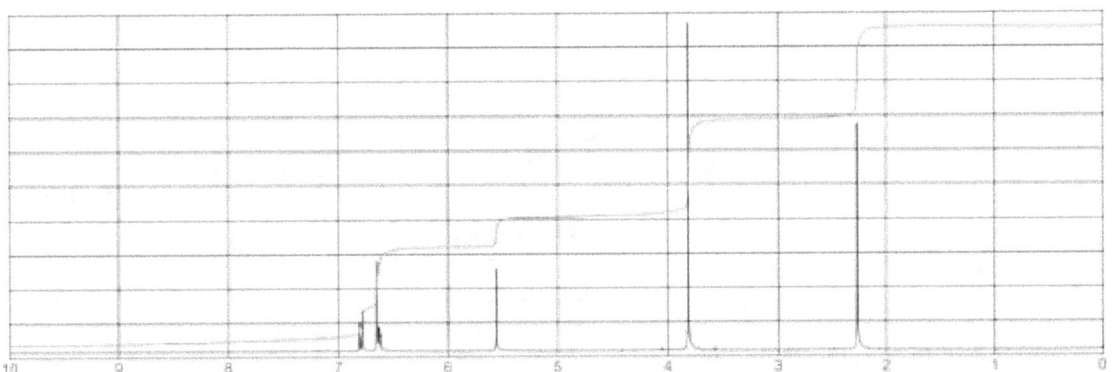

We can assign the four multiplets as follows:

Chemical shift δ (ppm)	Assignment(s)	Integral	Multiplicity
6.8	Aromatic hydrogen atoms	1	Doublet of doublets
6.6	Aromatic hydrogen atoms	2	Doublet of doublets
5.55	Ar – OH	1	Singlet
3.8	Ar – OCH_3	3	Singlet
2.3	Ar – OCH_3	3	Singlet

where Ar denotes the aromatic ring.

Whilst the last three peaks provide supporting evidence for the structure we can only be confident which candidate molecule is correct by examining the aromatic region.

Creosol

As already stated, the only way to determine the exact structure is to examine the aromatic hydrogen signal which is shown, in expanded form, below:

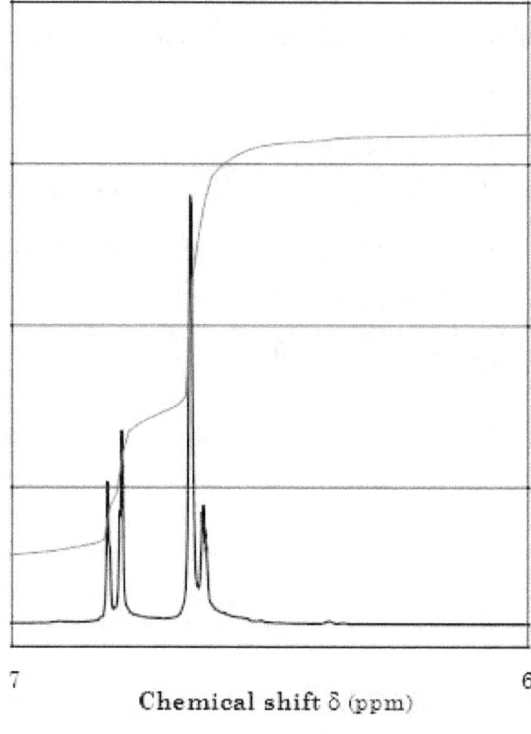

Chemical shift δ (ppm)

Observations

There are a number of matters to note:-

▪ There are two multiplets in the ratio of chemical shifts 1:2.

If we review the structures of the candidates again, we can see that candidates I, IV and VII would have the three hydrogen atoms are in very similar chemical and magnetic equivalent and so we can disregard them.

Creosol

This reduces the possible structures to the following candidates:

We must now make use of the coupling constants and the ^{1}H nmr spectrum tells us that there are three of value J = 9, 3 and 1 Hz. This immediately tells us that the three hydrogen atoms are positioned such that there are ortho– , meta– and para– arrangements and supports the supposition that the correct structure is one of the four shown above.

If we look at the spectra in even more detail we notice that the multiplets are in the ratio, in terms of chemical shift, 1:2.

* The peak at δ 6.8 ppm, of integral one, is deshielded and this must be due to its proximity to the phenolic oxygen atom as that is close to the multiplets in the ^{1}H nmr spectrum of

* The doublet of doublets at δ 6.6 ppm, of integral two, indicates that the two hydrogen atoms are, although not chemically and magnetically equivalent, existing in extremely similar environments. That they are of lower chemical shift than the multiple of integral one, indicates shielding and this may occur due to the inductive, electron donating effect, of the methyl group.

This leads us to the conclusion that only one of the possible structures, Candidate II, is a feasible structure as shown below:

We can find supporting information from the ^{13}C nmr spectrum which should, from the data sheet, contain six peaks in the region δ 160 – 160 ppm (due to the six aromatic carbon atoms), a peak in the region δ 50 – 90 ppm (due to the – OCH$_3$ carbon atom) and one peak in the region δ 0 – 50 ppm (due to the methyl carbon atom).

We consider the ^{13}C nmr spectrum next.

Creosol

^{13}C NMR Spectrum

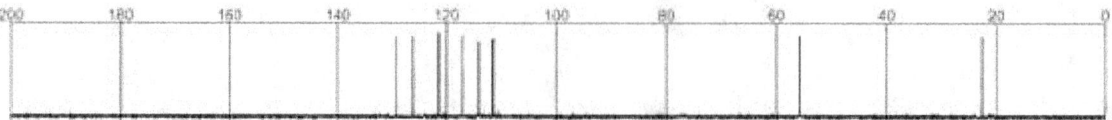

The ^{13}C nmr spectrum confirms the prediction and so we can be confident in the structure and can draw our final conclusions.

Observations and Conclusions

Structure

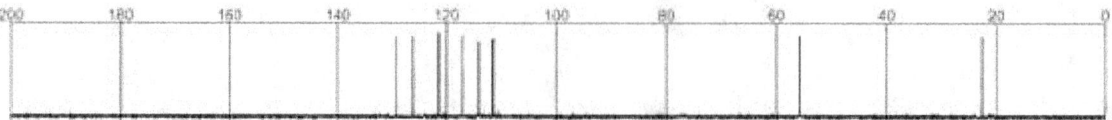

Systematic name: 2-methoxy-4-methylphenol

Chapter VII

Apocynin

Apocynin is a plant pheromone released by many plants when cut or wounded.

It is a fascinating compound as it has also been shown to have anti-inflammatory properties and is included in some asthmatic treatments as well as showing promise in treating arthritis, inflammatory bowel disease (IBD) and atherosclerosis which is one of the most common cardiovascular diseases in developed countries

Apocynin is a colourless, crystalline solid with a melting point of 115ºC and a boiling point of 335ºC and is almost completely insoluble in water.

This compound has the **elemental composition**: C: 65.01%, H: 6.08%, O: 28.89% and has the **formula mass (M_r)** of 166.12 g mol^{-1}.

These means that the empirical and molecular formulas are both $C_9H_{10}O_3$.

Again we have a molecule which may be a long chain aliphatic or aromatic compound, due to the number of carbon atoms, and, because of the presence of three oxygen atoms may be a carboxylic acid or an ester

Apocynin

Infrared Spectrum

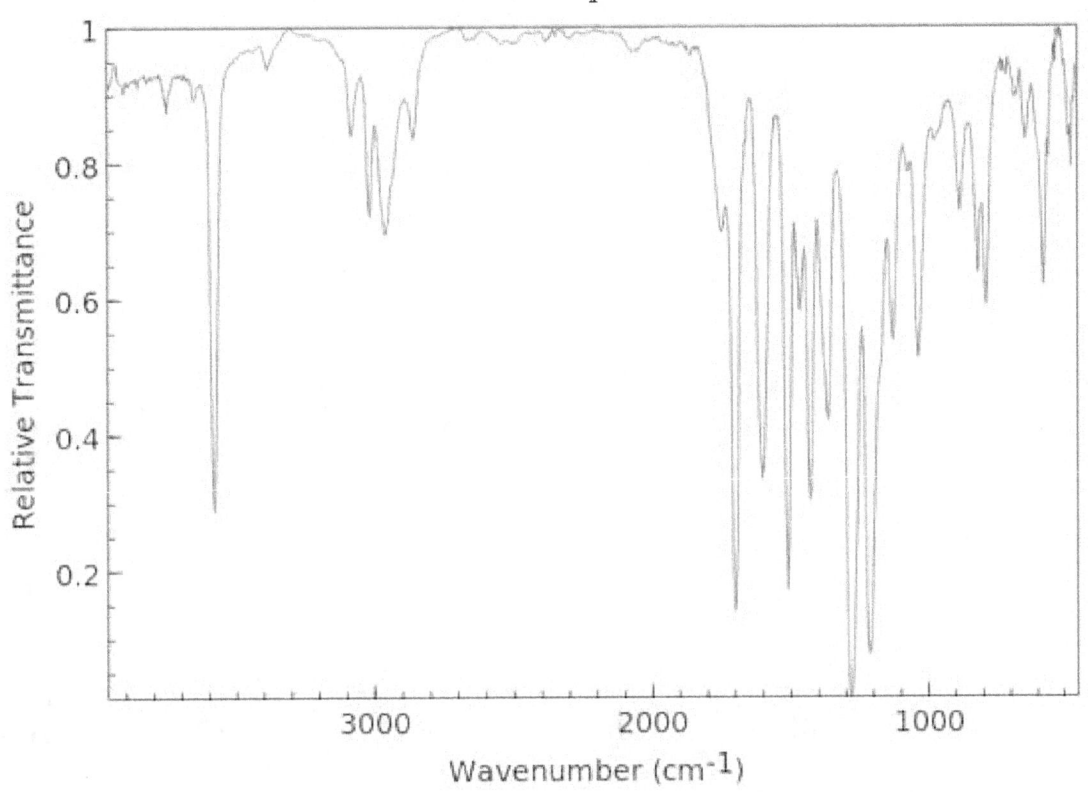

Observations

($\sqrt{}$ / X)	Wavenumber range (cm⁻¹)	Wavenumber (cm⁻¹)	Assignment
$\sqrt{}$	3200 - 3700	3580	**O – H**
X	3200 - 3600		**N – H**
$\sqrt{}$	3000 – 3300	3050	**C – H (aromatic)**
$\sqrt{}$	2500 – 3000	2995, 2910, 2805	**C – H (aliphatic)**
X	2200 – 2500		**C ≡ N**
$\sqrt{}$	1700 – 1800	1700	**C = O**
X	1600 – 1700		**C = C (aliphatic)**
$\sqrt{}$	1585 – 1600	1590	**C – C (aromatic)**
$\sqrt{}$	1450 – 1600	1500	**C – C (aromatic)**
$\sqrt{}$	1000 – 1300	1920	**C – O**
X	700 – 1000		**C – X** (X = Cl, Br or I)

Conclusions

This molecule contains an aromatic ring, an aliphatic substituent, a C = O and a C – O bond so must be an aromatic ester. The very sharp peak at 3580 cm⁻¹ is also characteristic of a phenol compound.

Apocynin

Mass Spectrum

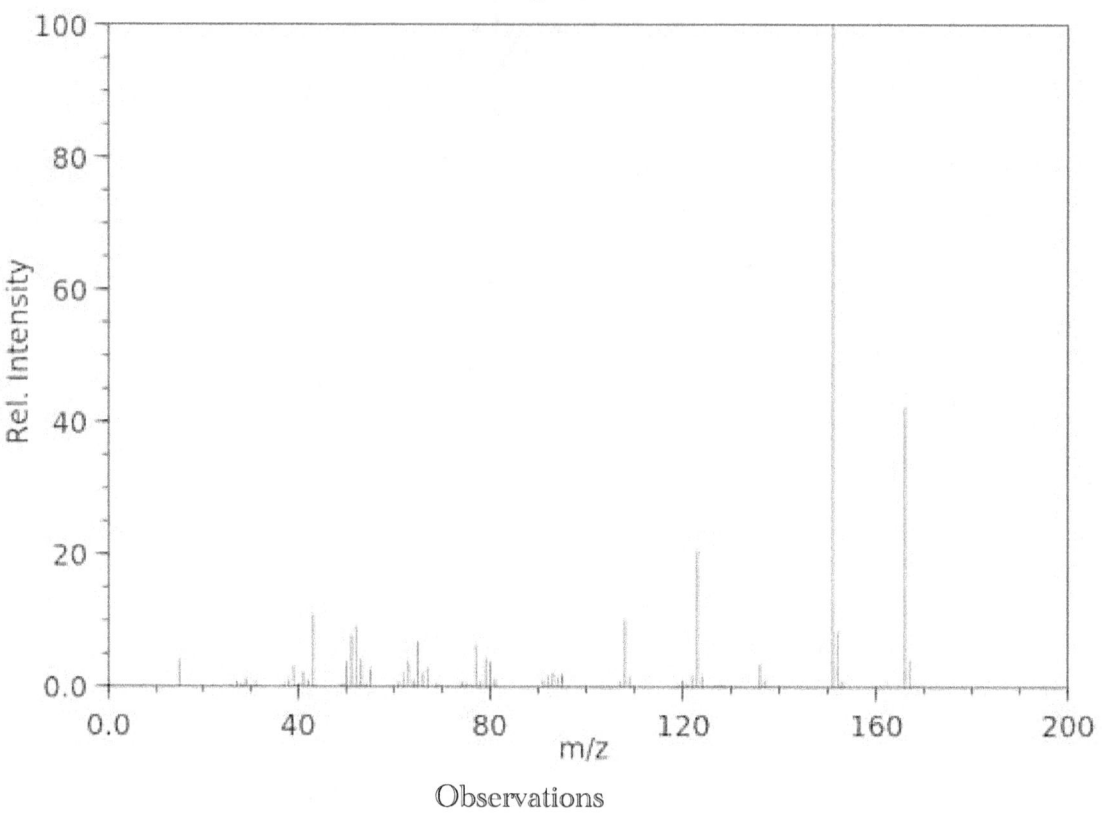

Observations

Charged fragments (m/z)	Assignment	Charged fragments (m/z)	Assignment
Molecular ion: 169	$[C_9H_{10}O_3]^+$	Base peak: 151	$[C_8H_7O_3]^+$
123	$[C_7H_7O_2]^+$	65	$[C_5H_5]^+$
108	$[C_6H_4O_2]^+$	52	$[C_4H_4]^+$
77	$[C_6H_5]^+$	43	$[C_2H_3O]^+$

Conclusions

- The peak at m/z = 77 indicates the presence of a benzene ring whilst,

- The peaks at m/z = 123 and 108 indicate the presence of, respectively, an ester grouping and an alkoxy group.

The presence of the ester grouping and the aromatic ring is also supported by the infrared spectrum. We can learn much more from the ^{1}H and ^{13}C nmr spectra which we consider next.

Apocynin

NMR Spectra

We will consider the spectra in turn and, for a change we will consider the ^{13}C nmr spectrum first.

^{13}C NMR Spectrum

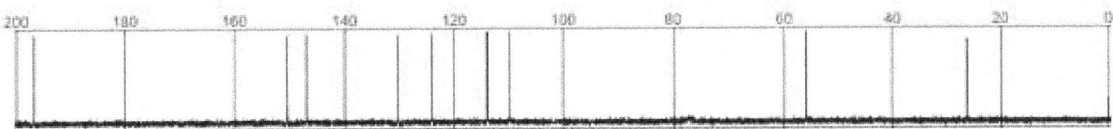

Using the data sheet, we can tabulate the signals as follows:

Chemical shift δ (ppm)	Assignment(s)	Integral
197	C = O	1
150	Ar – C	1
146	Ar – C	1
131	Ar – C	1
124	Ar – C	1
114	Ar – C	1
110	Ar – C	1
56	C – O	1
25	C – C	1

where *Ar* denotes the aromatic ring.

This conclusively demonstrates the existence of a benzene ring and implies that there is an ester grouping however the presence of the peak at δ 56 ppm suggests the presence of another grouping containing an Ar – O – CH₃ group.

This suggests that the molecule contains:-

- A six membered aromatic (benzene) ring;
- A group containing a C = O bond
- Another group containing an *Ar* – O – R group where R is an alkyl group.

This suggests that we can account for six of the carbon atoms in the ring, one in the alkoxy (*Ar* – O – R) group. That leaves us two account for and these may well be in an ester group.

The carbonyl group cannot be alone and simply attached to the ring and, since there is also an – OH group to attach, this suggests four potential candidates as shown below:

I	II	III	IV

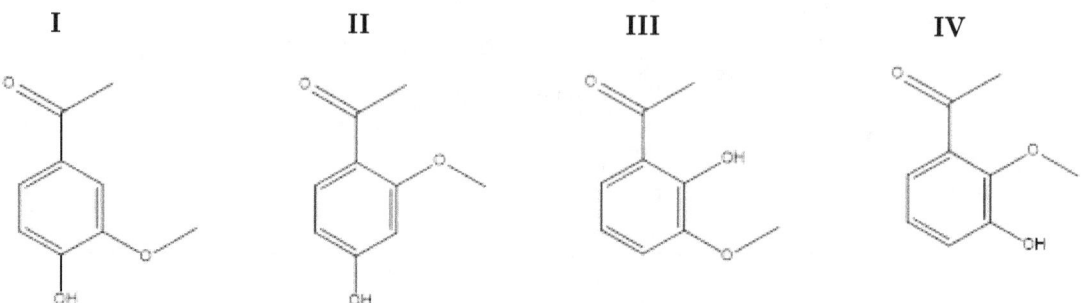

We can dispose of the ^{13}C nmr spectrum as it is no real help since *all* isomers will produce the following signals:

- Six in the aromatic region, δ 110 – 160 ppm;
- One due to the carbonyl, C = O, region, δ 160 – 220 ppm;
- One in the alkoxy region, δ 50 – 90 ppm;
- One in the alkyl region, δ 0 – 50 ppm

The ^{13}C nmr spectrum, shown below, contains all those peaks and, whilst it confirms that we are on the right track, it does not help in distinguishing between the isomers:

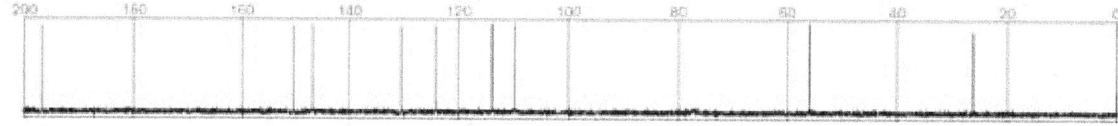

This leaves us to consider the ^{1}H nmr spectrum.

The ^{1}H nmr spectrum will contain signals due to the:

- Methyl group, – CH$_3$, attached to the carbonyl bond of integral three and will be a singlet;
- Methoxy group, – O – CH$_3$, which will also be a singlet of integral three.
- Three hydrogen atoms bonded directly to the benzene ring.

The three hydrogen atoms bonded to the aromatic ring are the key to determining the actual structure since we can predict the orientation of ortho –, meta – and para – hydrogen atoms and we can use the coupling constants to distinguish between them and identify the actual isomer which forms this pheromone. This is our next task and we will start by examining the 1H nmr spectrum and the expanded aromatic region

Apocynin

¹H NMR Spectrum

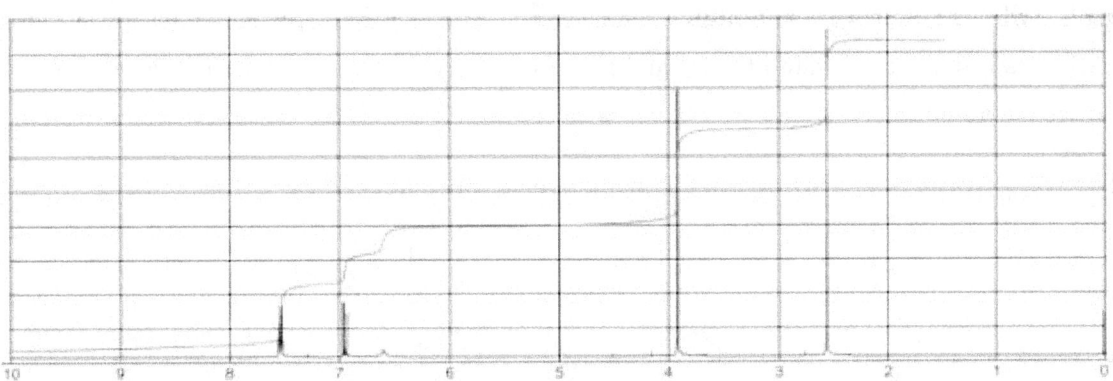

Observations

We can dispose of a couple of signals immediately:

- The singlet at δ 2.6 ppm of integral three must be due to the methyl, – CH_3, group;
- The singlet at δ 3.95 ppm of integral three must be due to the methoxy, – OCH_3 group;

If we examine the expanded δ 6 – 8 ppm region we can assign the peak at δ 6.5 ppm to the phenolic hydrogen atom and this leaves us to examine the aromatic region and the coupling constants of the multiplets.

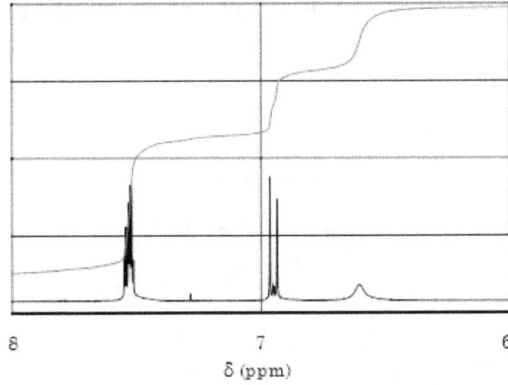

δ (ppm)

There are two multiplets to assign and we need to also examine the coupling constants which is our next and final task. The aromatic multiplets are as follows:

Chemical shift δ (ppm)	Multiplet	Integral	Coupling constants J (Hz)
7.55	Two overlapping doublets of doublets	2	*Multiplet 1*: 7 and 2 *Multiplet 2*: 2 and 1
6.9	Doublet of doublets	1	9 and 1

Apocynin

The fact that there are coupling constants demonstrating the existence of ortho –, meta – and para – hydrogen atoms immediately demonstrates that the only plausible structures are Candidates I and II as Candidates III and IV only contain ortho – and meta – arrangements of hydrogen atoms.

This means we are now down to the following two possible structures:

I II

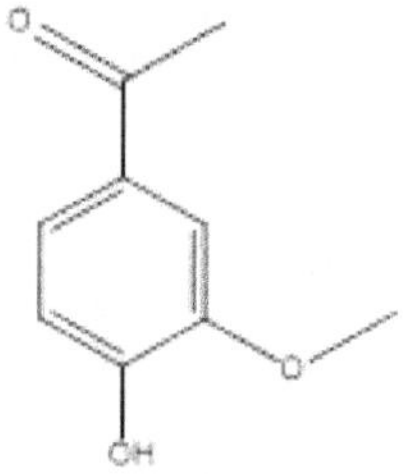

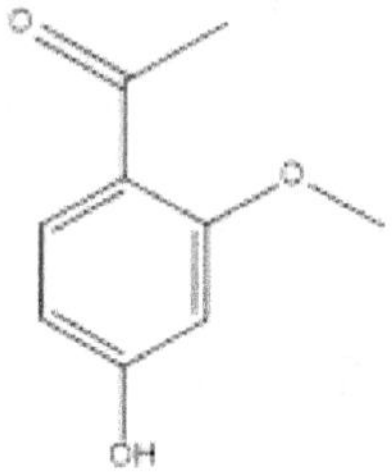

The ^{1}H nmr spectrum suggests that the molecule contains two hydrogen atoms of very similar chemical and magnetic environments and one in a distinctly different chemical and magnetic environment.

Only Candidate I has two hydrogen atoms which are in similar chemical and magnetic environments as indicated by the rectangle superimposed on the structure as shown below:

I

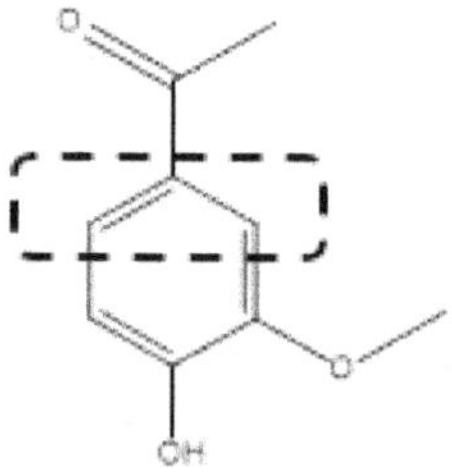

Observations and Conclusions

Structure:

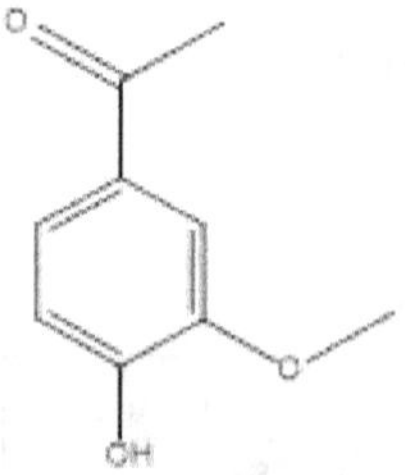

Systematic name: 1-(4-hydroxy-3-methoxyphenyl)ethan-1-one

Chapter VIII

Styron

Styron is another example of a pheromone which attracts and provides route directions for termites (pictured above) and which are closely related to cockroaches (pictured below).

A colourless crystalline solid with melting point 33ºC and boiling point 250ºC, it is slightly soluble in water and has the following properties:

Elemental composition: C: 80.48%, H: 7.53%; O: 11.92%

Formula mass (M_r): 134.2 g mol^{-1}.

This means that the empirical and molecular formulas are both $C_9H_{10}O$.

Styron

Infrared Spectrum

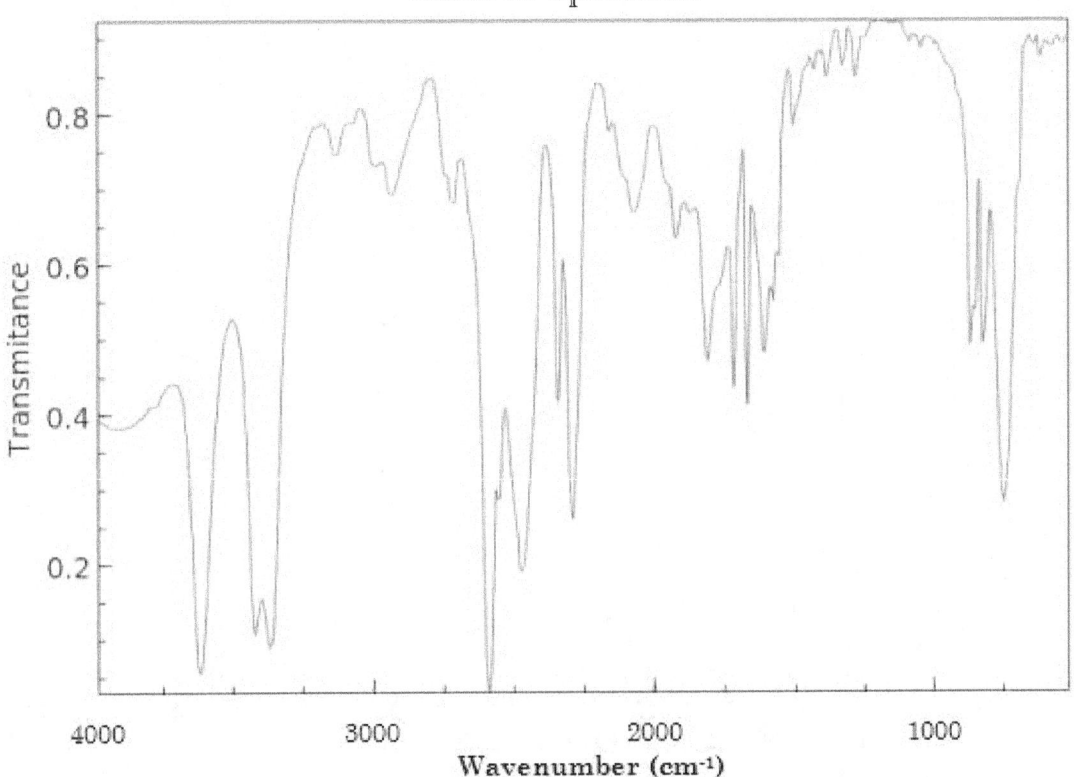

Wavenumber (cm⁻¹)

Observations

(√ / X)	Wavenumber range (cm⁻¹)	Wavenumber (cm⁻¹)	Assignment
√	3200 - 3700	3625	O – H
X	3200 - 3600		N – H
√	3000 – 3300	3390, 3325	C – H (aromatic)
√	2500 – 3000	2900	C – H (aliphatic)
X	2200 – 2500		C ≡ N
X	1700 – 1800		C = O
√	1600 – 1700	1625	C = C (aliphatic)
X	1585 – 1600		C – C (aromatic)
X	1450 – 1600		C – C (aromatic)
X	1000 – 1300		C – O
X	700 – 1000		C – X (X = Cl, Br or I)

Conclusions

This is a fascinating spectrum and, although many of the peaks cannot be assigned from the data sheet, it suggests that the molecule is aromatic and contains at least one alkyl group and at least one C = C bond.

Mass Spectrum

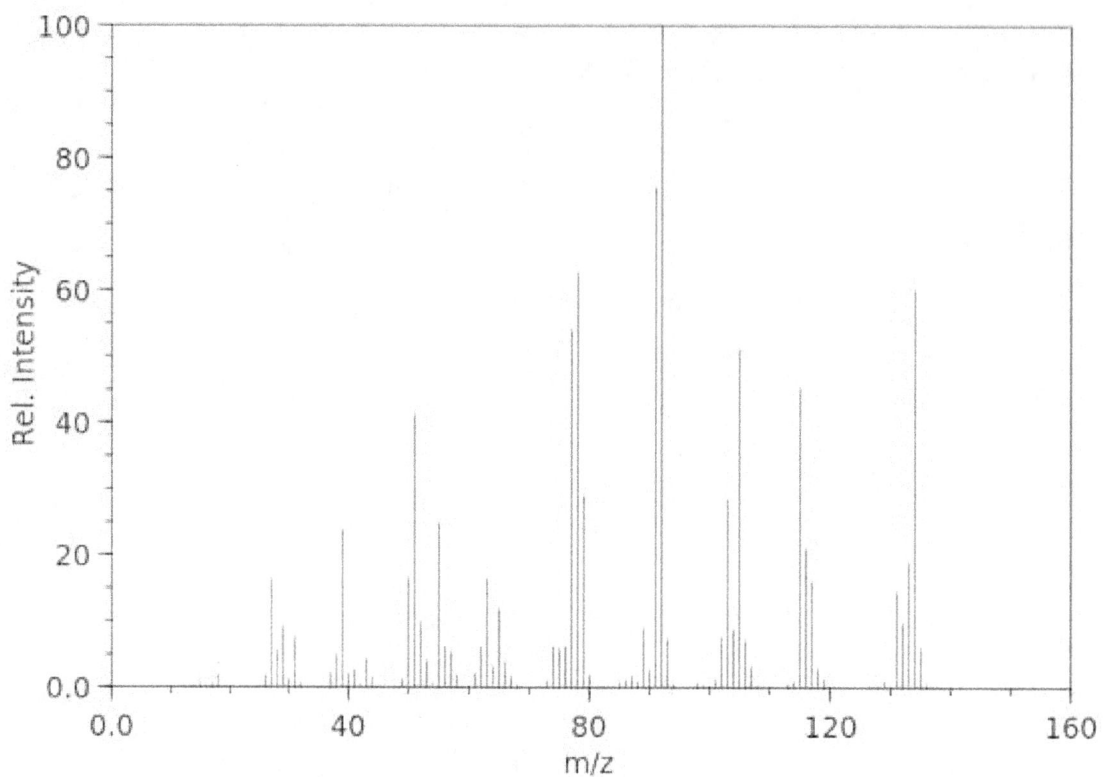

Observations

Charged fragments (m/z)	Assignment	Charged fragments (m/z)	Assignment
Molecular ion: 134	$[C_9H_{10}O]^+$	**Base peak: 92**	$[C_6H_5 - CH_2 + H]+$
115	$[C_8H_{10}O]^+$	51	$[C_4H_3]^+$
77	$[C_6H_5]^+$	39	$[C_3H_3]^+$
63	$[C_5H_3]^+$	27	$[C_2H_3]^+$

Conclusions

This spectrum is extremely detailed and many of the peaks have a range of possible assignments however the peaks at m/z = 77 indicates the presence of a mono-substituted benzene ring, C_6H_5-, with one substituent which is not hydrogen.

The peak at m/z = 39 suggests the presence of an allylic, CH – CH = CH –, group.

We can learn much more from the nmr spectra which we consider next, starting with the ^{13}C nmr spectrum.

NMR Spectra

^{13}C NMR Spectrum

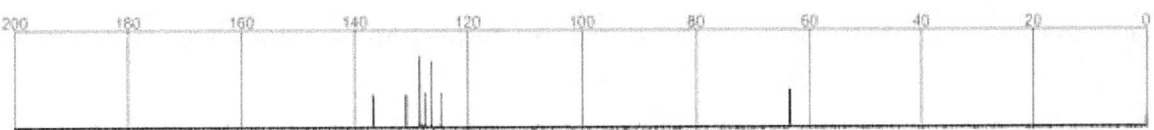

This spectrum is fascinating since there appear to be too many carbon atoms in the aromatic, δ 110 – 160 ppm. Whilst an aromatic ring can contain eight carbon atoms, the molecular formula prevents this as there are insufficient hydrogen atoms to fit in and so, at most, there must be a substituted benzene ring. We know, from the infrared spectrum, that the molecule contains a C = C bond and so we can account for the number of peaks by predicting the presence of two signals due to the alkene carbon atoms.

The peak at δ 63 ppm can be assigned to a – C – O – bond and we can learn more by examining the δ 110 – 160 ppm in detail.

There are two peaks of integral two, at δ 129 and δ125 ppm, and four peaks, all of integral one, at δ 137, 131, 128 and 125 ppm.

The two peaks of integral two cannot be due to the C = C carbon atoms and must be due to aromatic carbon atoms. That there are two such signals implies that the base structure of the molecule must contain two pairs of chemically and magnetically carbon atoms and so must start with

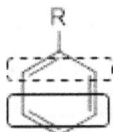

where the pairs of chemically and magnetically mutually equivalent carbon atoms are indicated by the solid and dashed lines and the remainder of the molecule is indicated by R.

One of the other aromatic signals must be due to the carbon atom para – to the functional group, R and the remainder of the molecule must contain three carbon atoms together with a C = C bond and a – C – OH group.

Styron

This group can only be of the following structure:

and so we can tentatively assign the following structure to the molecule.

which complies with not only the ^{13}C nmr spectrum but also accounts for many signals in the infrared and mass spectra.

This leaves us with the ^{1}H nmr spectrum which we can predict before examining the observed spectrum.

We do this now.

^{1}H NMR Spectrum

If we are correct with this structure, the ^{1}H nmr spectrum will contain the following signals.

- A signal to the – **OH** hydrogen atom which will be of integral one but can appear anywhere in the spectrum and which will be a singlet;
- A doublet of integral two due to the – **CH2** – OH group which will appear in the δ 3 – 4.2 ppm. The doublet will be caused by the closest alkene hydrogen atom;
- If we consider the C = C double bond hydrogens we can immediately suggest that they will produce signals in the δ 4.5 – 6 ppm. Both will be of integral one but:-
    - One will be a doublet of triplets due to its proximity to the – CH2 – group and the other C = C hydrogen atom as highlighted below:

    The signal will be split into a triplet by the – CH2 – but this will then be split into a doublet of triplets by the other hydrogen atom on the C = C bond.

■ The second alkene, **CH** = CH hydrogen atom will be a doublet since there is only one hydrogen atom on adjacent carbon atoms, again, as shown below:

This leaves us with the aromatic hydrogen atoms to consider and to determine our predictions we will use the following labelled structure:

We will consider H(a), H(b) and H(c). H(a') and H(b') will behave exactly the same as H(a) and H(b) respectively which is why H(a) / H(a') & H(b) / H(b') would both give rise to multiplets of integral two.

■ Starting with H(a)

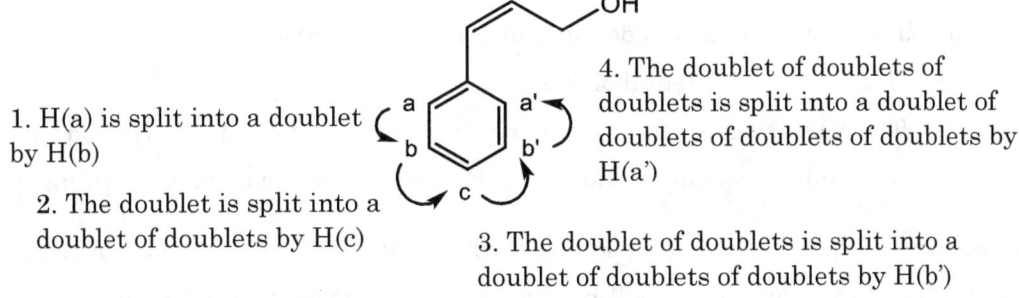

1. H(a) is split into a doublet by H(b)

2. The doublet is split into a doublet of doublets by H(c)

4. The doublet of doublets of doublets is split into a doublet of doublets of doublets of doublets by H(a')

3. The doublet of doublets is split into a doublet of doublets of doublets by H(b')

There will, therefore be a doublet of doublets of doublets of doublets in the aromatic region and, H(a) is ortho – to H(b), meta – to H(c) and H(a') and para – to H(b') we will observe coupling constants corresponding to all three of the ranges of coupling constants recorded in the data sheet.

We continue with H(b) and then H(c).

* With regard to H(b)

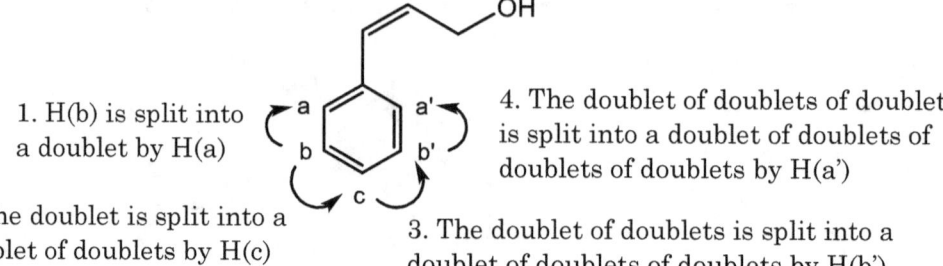

1. H(b) is split into a doublet by H(a)

2. The doublet is split into a doublet of doublets by H(c)

4. The doublet of doublets of doublets is split into a doublet of doublets of doublets of doublets by H(a')

3. The doublet of doublets is split into a doublet of doublets of doublets by H(b')

H(b) is ortho – to H(a) and H(c), meta – to H(b') and para – to H(a') so again we will observe coupling constants relating to all three ranges.

* For H(c)

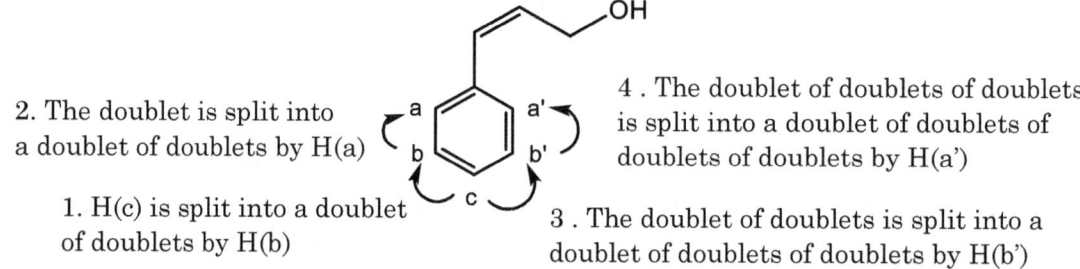

2. The doublet is split into a doublet of doublets by H(a)

1. H(c) is split into a doublet of doublets by H(b)

4 . The doublet of doublets of doublets is split into a doublet of doublets of doublets of doublets by H(a')

3 . The doublet of doublets is split into a doublet of doublets of doublets by H(b')

H(c) is ortho – to H(b) and H(b'), meta – to H(a). There are no para – arrangements.

This means that we can distinguish between the peaks in several ways:

* H(c) will produce a doublet of doublets of doublets of doublets but:
    * It will have an integral of one and;
    * The coupling constants will be within the range of J = 2 – 10 Hz and there will no coupling constants below J = 2 Hz since there are no para – arrangements.

We can now look back at the alkene peaks which will comprise a doublet of triplets and a doublet. The C = C hydrogens will be capable of forming an E – (**entgegen** – *against* – or *trans*) or a Z – (**zusammen** – *together* – *cis*) configuration and we can distinguish between them on the basis of their coupling constants as well.

We will consider the observed ¹H nmr spectrum next.

^{1}H NMR Spectrum

Observations and Conclusions

The ^{1}H nmr spectrum is shown below:

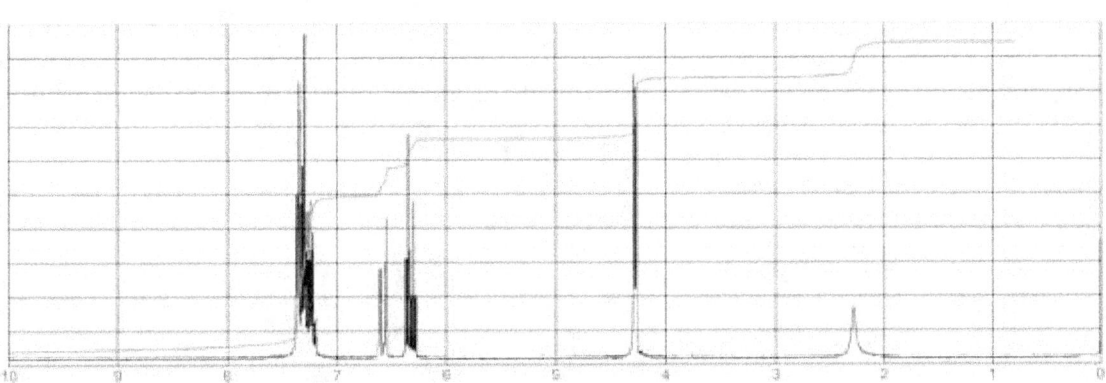

Two of the peaks are easy to assign and dispose of:

* That at δ 2.3 ppm, of integral one, must be due to the hydroxyl, – OH, hydrogen atom.

* The doublet, of integral two, at δ 4.3 ppm occurs in the – O – C region so we can assign it to – CH_2 – OH.

This leaves us to examine the δ 6 – 8 ppm region in more detail and the expanded spectrum is shown below:

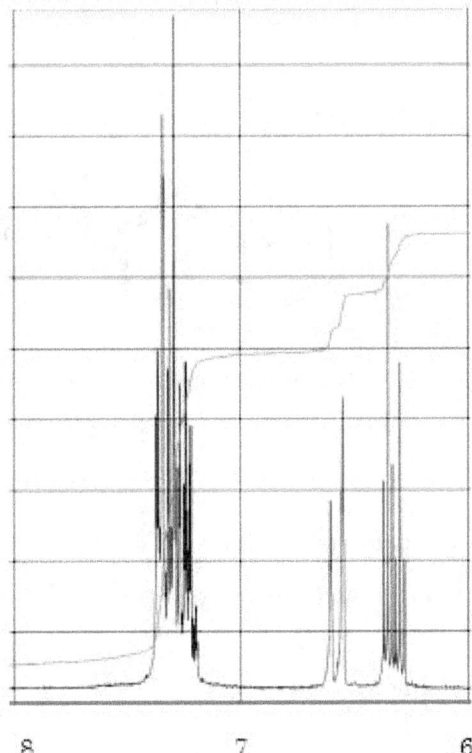

We will examine the alkene region first and the expanded region is shown below:

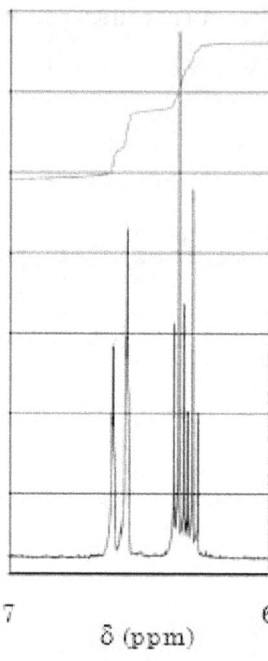

δ (ppm)

▪ The multiplet at δ 6.35 ppm, of integral one, is a doublet of triplets with coupling constants of J = 17 Hz and J = 7 Hz. We can label this as H(d).

▪ The doublet at δ 6.6 ppm, of integral one, with one coupling constant of J = 17 Hz. We can label this as H(e).

We can assign these peaks as follows

▪ H(d) is a doublet, of integral one due to splitting by H(e);

▪ H(e) is a doublet of triplets due to splitting by the two hydrogen atoms, H(f)
  – causing a triplet – which is then split in to a doublet of triplets by H(d).

H(f) has already been accounted for (the doublet δ 4.3 ppm) and this simply leaves us to account for the aromatic signals which comprise multiplets of integral five.

This is our penultimate task.

The actual aromatic spectrum comprises two multiplets of integral two and another one of integral one as shown below:

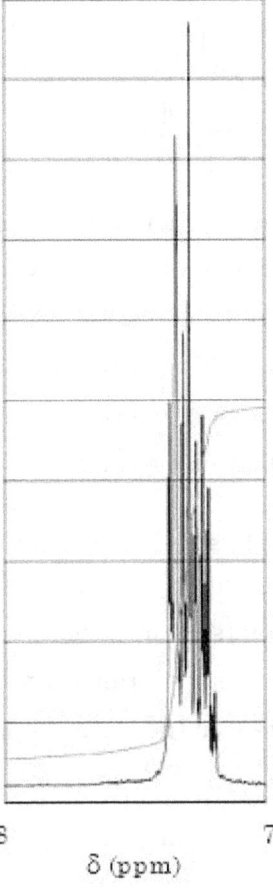

δ (ppm)

Whilst initially thinking this complex overlapping and daunting to analyse, it is less complex than it appears:

- There are two doublets of doublets of doublets of doublets which overlap but have the following coupling constants:
    - Multiplet 1: J = 8, 7, 2 and 1 Hz
    - Multiplet 2: J = 8, 3, 2 and 1 Hz

This means that:
    - One multiplet reports the existence of two ortho – , one meta – and one para – arrangements of the aromatic hydrogen atoms whilst;
    - The second multiplet identifies the occurrence of one ortho – , two meta – and one para – distributions.

If we examine the proposed structure again, below, we can identify the causes of both multiplets.

- With regard to H(a) we can see that H(a) is ortho – to H(b), meta – to H(a') and H(b) and para – to H(b'). The same applies if we work from H(a') which is why we have one doublet of doublets of doublets of doublets of integral two with:-
  - One coupling constant in the range J = 7 – 10 Hz (ortho –);
  - Two coupling constants in the range J = 2 – Hz (meta –) and;
  - One coupling constant in the range J = 0 – 2 Hz (para –).

- Considering from H(b) we can see that H(b) is ortho – to H(b) and H(c), meta – to H(b') and para – to H(a') and so we should expect :-
  - Two coupling constants in the range J = 7 – 10 Hz (ortho –);
  - One coupling constant in the range J = 2 – Hz (meta –) and;
  - One coupling constant in the range J = 0 – 2 Hz (para –).

This is exactly what the proposed structure predicts and this leaves us with examining the remaining multiplet which must be caused by H(c) as labelled in the structure above.

H(c) split into a triplet by H(b) and H(b') and this is split into a triplet of triplets by H(a) and H(a') as shown below where the initial splitting of the signal into a triplet by H(b) and H(b') is indicated by the solid arrows and the subsequent splitting of a triplet into a triplet of triplets by H(a) and H(a') is indicated by the dashed arrows.

This leaves us with one final task which is to return to the C = C bond and the orientation of the two hydrogen atoms.

Styron

We can now be confident that the overall structure of the molecule is correct but there are two possible orientations of the hydrogen atoms on the C = C bond.

They are shown below with a truncated – CH_2OH group attached to the C = C bond:

Isomer I

Since both hydrogen atoms are on the same side of the C = C bond this is described as a *zusammen* (German: *together*) isomer and is denoted as (Z –). In older terminology it would be described as *cis-* and although the latter terminology is no longer used it does crop up in older textbooks and in papers.

Isomer II

Since the hydrogen atoms are on opposite sides of the C = C bond this is described as the *entgegen* (German: *against*) isomer and is denoted as (E –). In older terminology it is described as *trans-*.

From the data sheet we can distinguish between E – and Z – isomers by measurement of the coupling constant since the coupling constants for E – isomers produce coupling constants in the range J = 15 – 20 Hz whilst the coupling constants for Z – isomers lie between J = 5 Hz and J = 14 Hz.

The observed spectrum demonstrates a coupling constant of J = 17 Hz and so this molecule is the E – isomer.

Conclusions

Structure:

Systematic name: (2E)-3-Phenylprop-2-en-1-ol.

This is a complicated name and other commonly used names for styron are cinnamyl alcohol and also phenylallyl alcohol.

Chapter IX

Linolenic acid

Linolenic acid is a *queen retinue pheromone* which leads the queen bee (pictured above) to attract a circle of bees (her retinue).

It also polymerises to form lignin which is a key component in the support structure of plants and wood. Esterification with fatty acids produces a waxy protective covering of apple.

Linolenic acid is a colourless liquid with a melting point of -21°C and a boiling point of 443°C and has the following properties:

Elemental composition: C: 77.59%, H: 10.88%; O: 11.49%

Formula mass (M_r): 278.4 g mol^{-1}.

This means that it has the following composition:-

Empirical formula is $C_9H_{15}O$

Molecular formula is $C_{18}H_{30}O_2$.

Linolenic acid

Infrared Spectrum

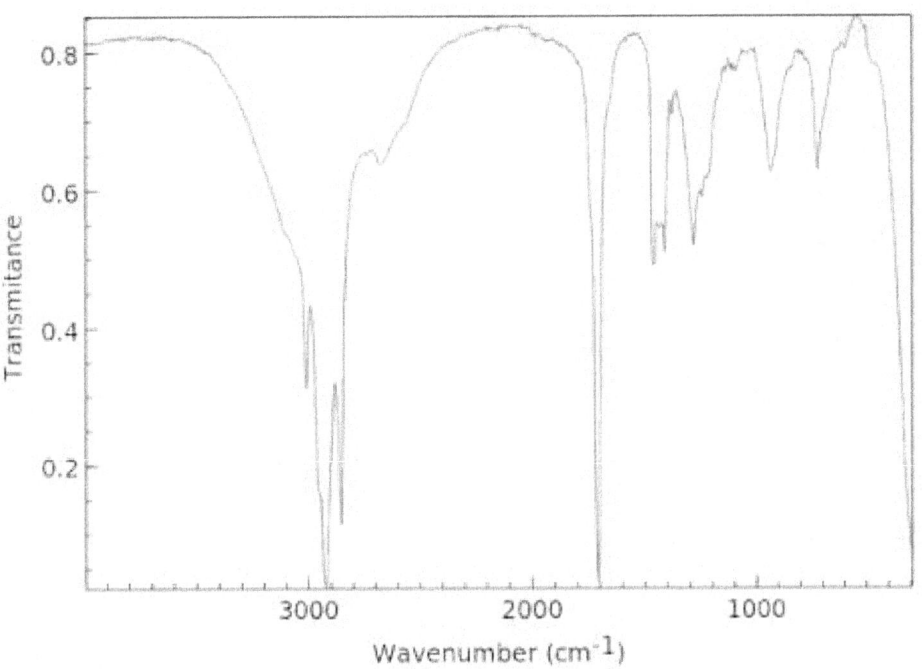

Observations

(√ / X)	Wavenumber range (cm⁻¹)	Wavenumber (cm⁻¹)	Assignment
√	3200 - 3700	Undefined	**O – H**
X	3200 - 3600		**N – H**
√	3000 – 3300	3010	**C – H (aromatic / unsaturated)**
√	2500 – 3000	2920, 2830	**C – H (aliphatic)**
X	2200 – 2500		**C ≡ N**
√	1700 – 1800	1705	**C = O**
X	1600 – 1700		**C = C (aliphatic)**
X	1585 – 1600		**C – C (aromatic)**
X	1450 – 1600		**C – C (aromatic)**
√	1000 – 1300	1250	**C – O**
X	700 – 1000		**C – X (X = Cl, Br or I)**

Conclusions

With aliphatic C – H bonds, a strong C = O bond and a possible but unusually shaped O – H peak as well as a weak unsaturated C – H bond and a C – O bond stretch it is likely that this compound is an unsaturated carboxylic acid and we can learn more from the mass spectrum.

Linolenic acid

Mass Spectrum

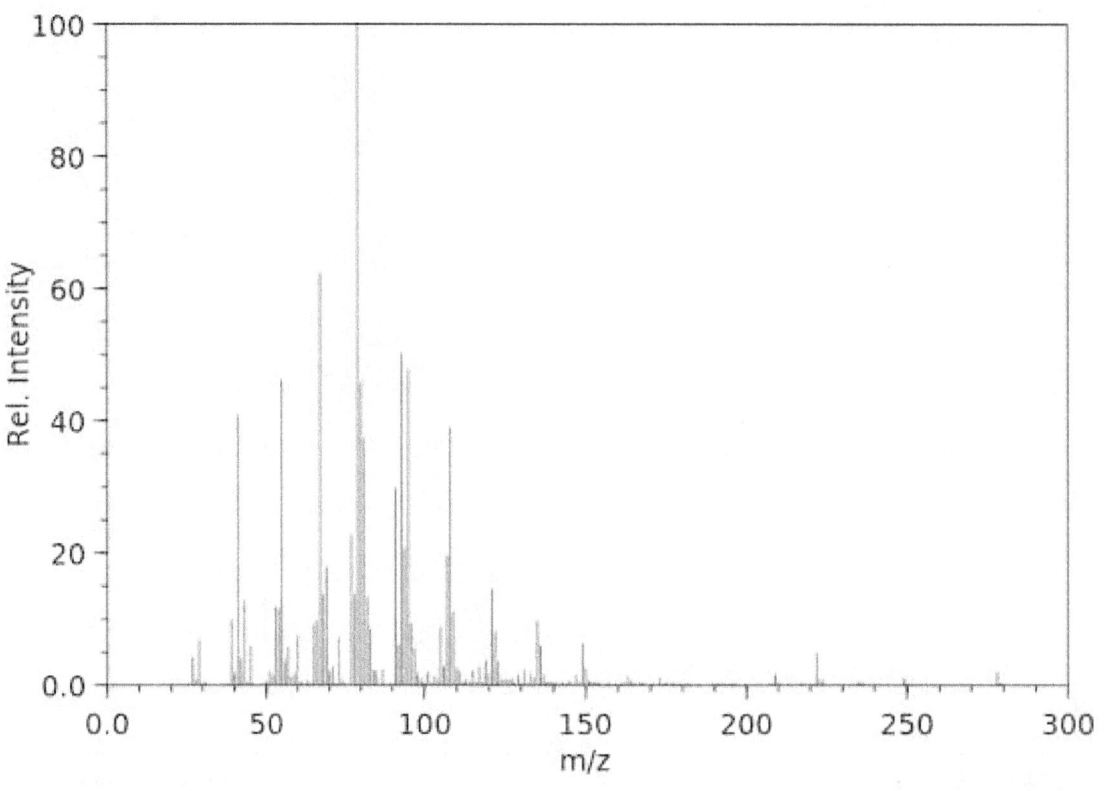

Observations

Charged fragments (m/z)	Assignment	Charged fragments (m/z)	Assignment
Molecular ion: 278	$[C_{18}H_{30}O_2]^+$	**Base peak: 79**	$[C_7H_7]^+$
149	$[C_{11}H_{17}]^+$	67	$[C_5H_7]^+$
135	$[C_{10}H_{15}]^+$	55	$[C_4H_7]^+$
121	$[C_9H_{13}]^+$	43	$[C_3H_7]^+$
108	$[C_8H_{12}]^+$	41	$[C_3H_5]^+$
95	$[C_7H_{11}]^+$	29	$[C_2H_5]^+$

Conclusions

This spectrum implies that the molecule contains a long chain hydrocarbon with at least one C = C bond which concurs with the infrared spectrum but provides no supporting evidence for the existence of an ester or carboxylic acid but it does support the existence of a substituted, aromatic, six – membered ring. We can learn much more from the ^{1}H and ^{13}C nmr spectra which we consider next.

Linolenic acid

NMR Spectra

The ^{1}H and ^{13}C nmr spectra are shown below and appear, initially at least, challenging.

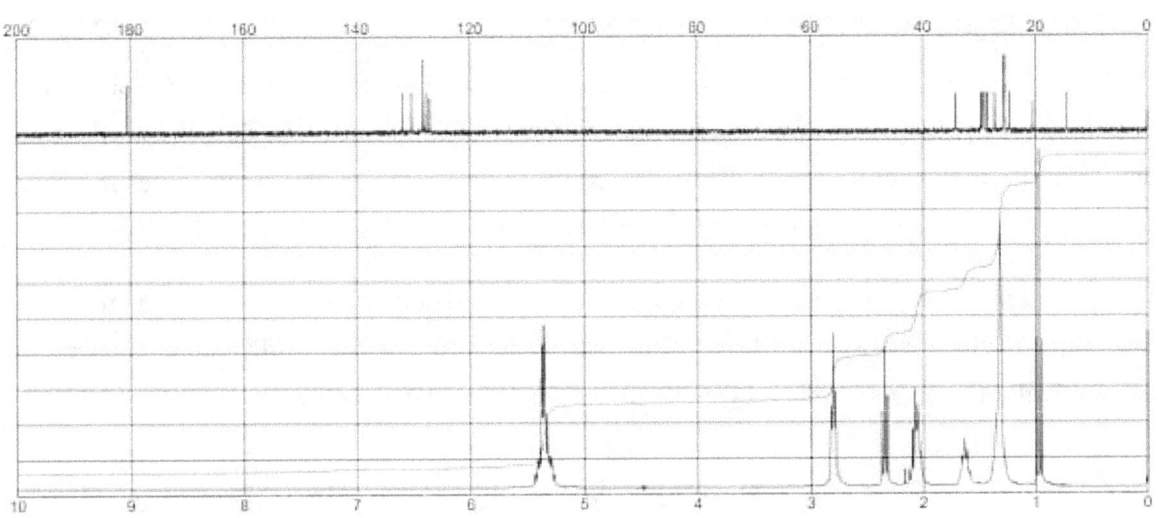

Analysing it is not, however, as daunting as it looks.

If we examine the ^{13}C nmr spectrum we can tabulate the peaks as shown below:

Chemical shift δ (ppm)	Integral	Assignment(s)
181	1	C = O
133	1	C = C
131	1	C = C
127	2	C = C
126	1	C = C
125	1	C = C
124	1	C = C
35	1	C − C
31	1	C − C
30	1	C − C
29	1	C − C
27	1	C − C
25	2	C − C
24	1	C − C
21	1	C − C
15	1	C − C

Linolenic acid

Observations and Conclusions

This spectrum is very significant.

The infrared spectrum implies the existence of at least one $C = O$ and at least one $C = C$ bond.

The ^{13}C nmr spectrum indicates the existence of one $C = O$ bond and three $C = C$ bonds and the latter are intriguing. There are six aromatic or $C = C$ atoms and two, the peak of integral two, have the same chemical shift. This indicates that four of the carbon atoms are chemically and magnetically *non-equivalent* but two **are** chemically and magnetically equivalent.

The peak of integral two must indicate that the $C = C$ bonds must have a $C = C$ bond on either side, sequentially, otherwise *all* the $C = C$ carbon atoms would be chemically and magnetically non-equivalent and this implies the existence of the following structure,

where the squiggle, ⌇ , indicates an undefined group.

This leaves us with the alkyl groups and the $C = O$ functional group to be placed on the molecule. Whilst there are a number of substituted isomers of this molecular formula which can be drawn there would be a number of chemically and magnetically equivalent carbon atoms and so there would a number of peaks in the alkyl region of integral of at least two. In fact, there is only one such peak.

The simplest structure is linear as shown below:

The position of the $C = C$ bonds can, however, be at any place in the chain other than in the terminal positions and so a number of isomers can exist.

There is also the matter of the orientation of the $C = C$ bonds each of which can be an E – or Z – isomer. This means that there are an even larger number of possible isomers. It is pointless to draw them all out and predict the ^1H nmr spectrum of each one and it is far more efficient and less headache – inducing to examine the ^1H nmr spectrum which is our final task.

Linolenic acid

¹H NMR Spectrum

The alkyl region of the ¹H nmr spectrum is shown below.

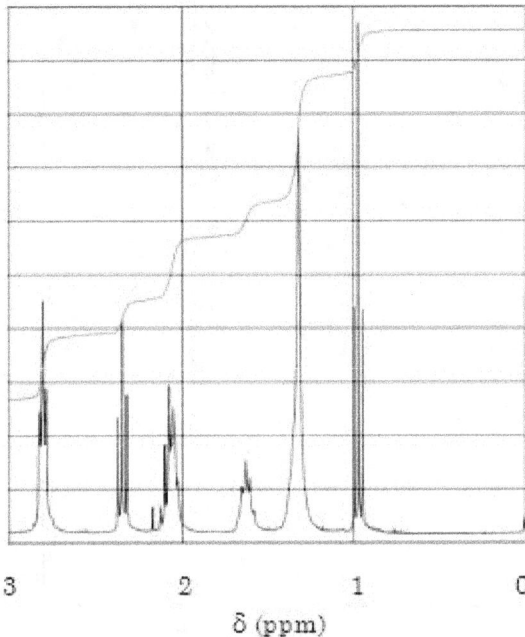

We can immediately discount the proposed structure as there is no doublet of integral three which would arise from the terminal $-$ CH_3 group bonded to a C = C bond. There is, however, a triplet of integral three at δ 0.95 ppm which can only arise from a $CH_3 - CH_2 -$ group (n+1 rule) suggesting that the molecule might have the structure shown below with the hydrogen atoms labelled with letters:

If this structure is correct then there will be :-

■ A triplet, of integral three, due to H(a) being split by H(b) and

■ A triplet, of integral two, due to H(q).

These appear at δ 0.95 ppm and δ 1.62 ppm respectively.

We can examine the remainder of the alkyl region but it is best to study the δ 5 – 6 ppm region first which we do now.

Linolenic acid

Examination of the alkene region

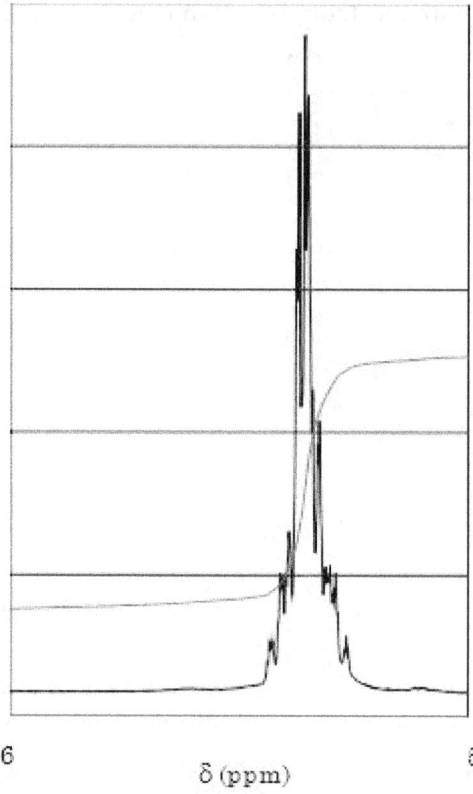

6

δ (ppm)

5

The multiplet centred on δ 5.35 ppm, of integral six, is a series of overlapping doublets of triplets.

- H(c) is split into a triplet by H(b) which is split into a doublet of triplets by H(d);
- H(d) is split into a triplet by H(e) which is split into a doublet of triplets by H(c);
- H(f) is split into a triplet by H(e) which is split into a doublet of triplets by H(g);
- H(g) is split into a triplet by H(h) which is split into a doublet of triplets by H(f);
- H(i) is split into a triplet by H(h) which is split into a doublet of triplets by H(j);
- H(j) is split into a triplet by H(k) which is split into a doublet of triplets by H(i);

This explains why there is a complex multiplet, of integral six, in the alkene region and there is the interesting matter of coupling constants which we will return to after examining the remaining alkyl signals which becomes a relatively simple matter and which is our next task.

Linolenic acid

Alkyl hydrogens

The expanded alkyl region is reproduced below again and we examine the regions highlighted.

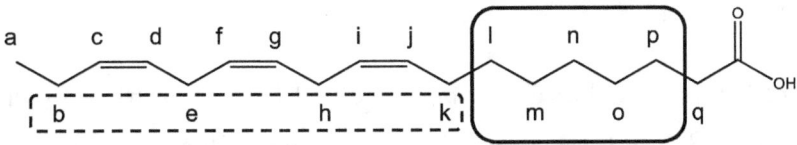

Predicting and then examining the ¹H nmr spectrum and considering the dashed rectangle highlighted hydrogen atoms first:-

- H(b) will produce a doublet of quartets, of integral two, since its signal is split by the three H(a) atoms and the resultant quartet is split into a doublet of quartets by H(c).
- H(e) will produce a triplet, of integral two, since it is split by H(d) and by H(f).
- H(h) will produce a triplet, of integral two, since it is split by H(g) and by H(i).
- H(k) will produce a doublet of triplets, of integral two, since its signal is split by the two H(l) atoms and the resultant triplet is split into a doublet of triplets by H(j).

Due to the similarity of their environments, the signals due to H(e) and H(h) will, at least, overlap and may appear in precisely the same position which means we will look for a complex multiplet of integral four.

Equally, H(b) and H(k) inhabit similar chemically and magnetically equivalent environments so we will be looking for another multiplet of integral four.

There are two such multiplets, each of integral four, at δ 2.8 ppm and δ 2.1 ppm and we can presume that the multiplet at δ 2.8 ppm is due to H(e) and H(h) as they are influenced by the C = C bonds whilst H(b) and H(k) are influenced by their proximities to alkyl groups.

This leaves us with the hydrogen atoms highlighted with the solid rectangle.

- H(l), of integral two, will produce a triplet due to H(k) which is split into a triplet of triplets by H(m);
- H(m), of integral two, will produce a triplet due to H(l) which is split into a triplet of triplets by H(n);
- H(n), of integral two, will produce a triplet due to H(m) which is split into a triplet of triplets by H(o);

Linolenic acid

- H(o), of integral two, will produce a triplet due to H(n) which is split into a triplet of triplets by H(p);

- H(p), of integral two, will produce a triplet due to H(o) which is split into a triplet of triplets by H(q);

These atoms account for the signal at δ 1.35 ppm and justifies the entire assignments even though there is no signal due to the hydroxyl hydrogen atom. This is very common and its existence is revealed by the ^{13}C nmr spectrum.

Our final task is to consider the orientation of the hydrogen atoms on the C = C bonds i.e. whether each is E – or Z –. The matter is easily resolved since all the coupling constants in the multiplet centred at δ 5.35 ppm are J = 7 Hz and we can, quite confidently, conclude the following:

Conclusions

Structure:

Systematic name: (9Z,12Z,15Z)-octadeca-9,12,15-trienoic acid

Chapter X

Banana oil

One of the major chemical components of bananas, ***banana oil*** is a colourless liquid which is sparingly soluble in water. Some people find the odour also resembles that of pears.

Dissolved in ethanol, banana oil is widely used in the food industry to create artificial banana flavourings and is also used as a solvent.

Released when a honey bee (pictured above) stings it also serves as a beacon pheromone to attract other bees and provoke them to also sting.

Sparingly soluble in water and with a melting point of -78°C and a boiling point of 142°C, this compound has the following specific properties:

Elemental composition: C: 64.52%, H: 10.86%; O: 24.58%
Formula mass (M_r): 130.2 g mol^{-1}.

These means that the empirical and molecular formulas are both $C_7H_{14}O_2$.

Banana oil

Infrared Spectrum

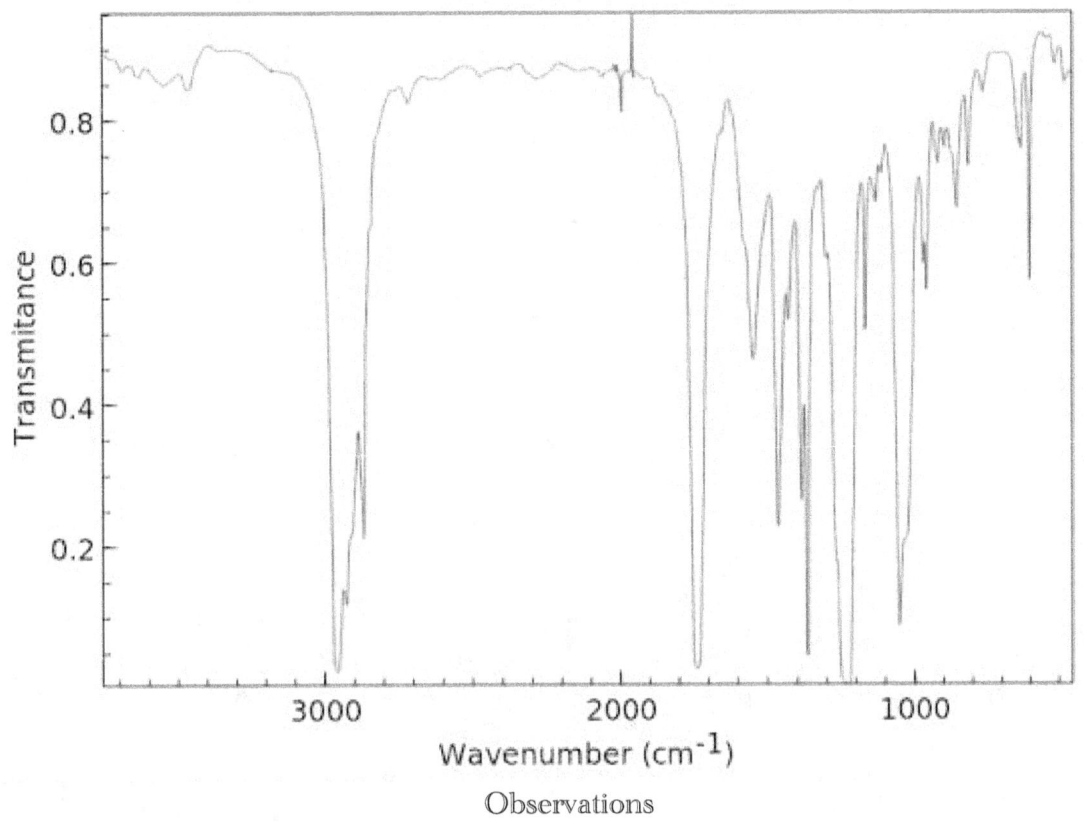

Observations

(√ / X)	Wavenumber range (cm⁻¹)	Wavenumber (cm⁻¹)	Assignment
X	3200 - 3700		O – H
X	3200 - 3600		N – H
X	3000 – 3300		C – H (aromatic)
√	2500 – 3000	2960, 2950, 2820	C – H (aliphatic)
X	2200 – 2500		C ≡ N
√	1700 – 1800	1740	C = O
X	1600 – 1700		C = C (aliphatic)
X	1585 – 1600		C – C (aromatic)
X	1450 – 1600		C – C (aromatic)
√	1000 – 1300	1230	C – O
X	700 – 1000		C – X (X = Cl, Br or I)

Conclusions

Due to the aliphatic C – H stretches and, the strong, C = O and C – O peaks this compound is clearly an aliphatic ester and we can learn more from the mass spectrum.

Banana oil

Mass Spectrum

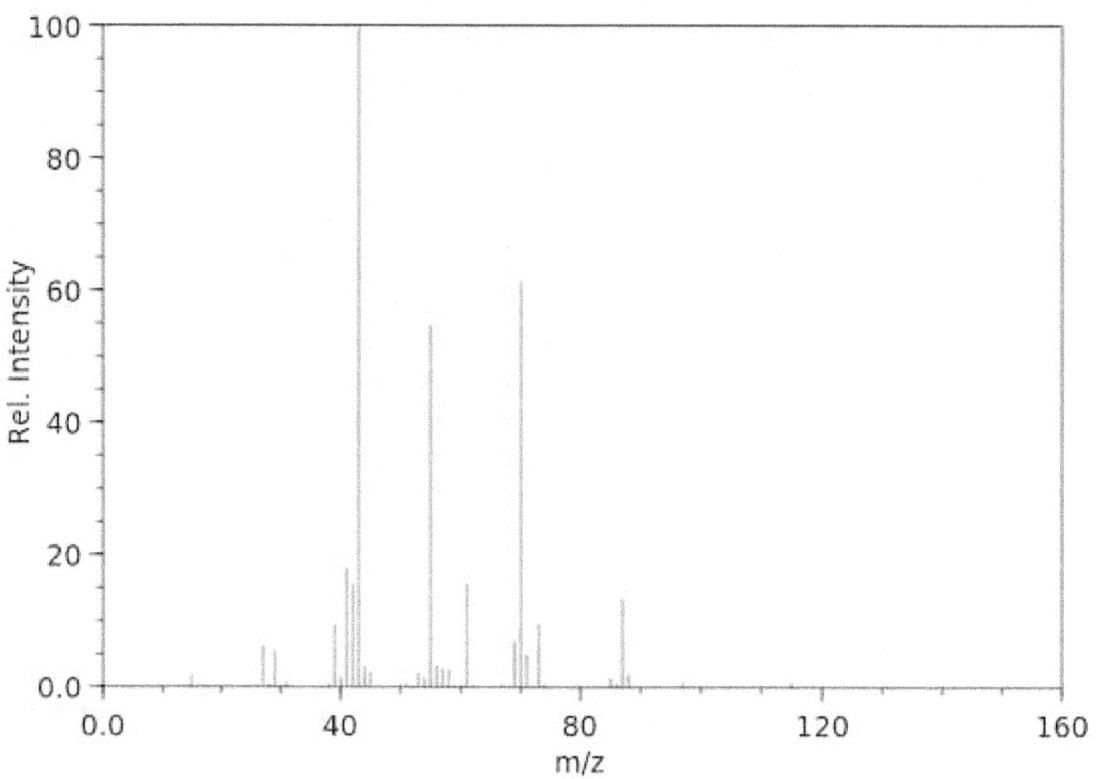

Observations

Charged fragments (m/z)	Assignment	Charged fragments (m/z)	Assignment
Molecular ion: 87	$[CH_2CO_2C_2H_5]^+$	Base peak: 43	$[CH_3CO]^+$
70	$[C_5H_{10}]^+$	39	$[C_3H_3]^+$
61	$[CH_3OCOH_2]^+$	29	$[C_2H_5]^+$
55	$[C_4H_7]^+$	15	$[CH_3]^+$

Conclusions

There is no evidence for the existence of an aromatic ring but the peaks do suggest and support the inference from the infrared spectrum that this is an aliphatic carboxylic ester.

Banana oil

NMR Spectra

The nmr spectra, displayed below, are pleasingly simple to analyse and we consider the ^{13}C nmr spectrum first.

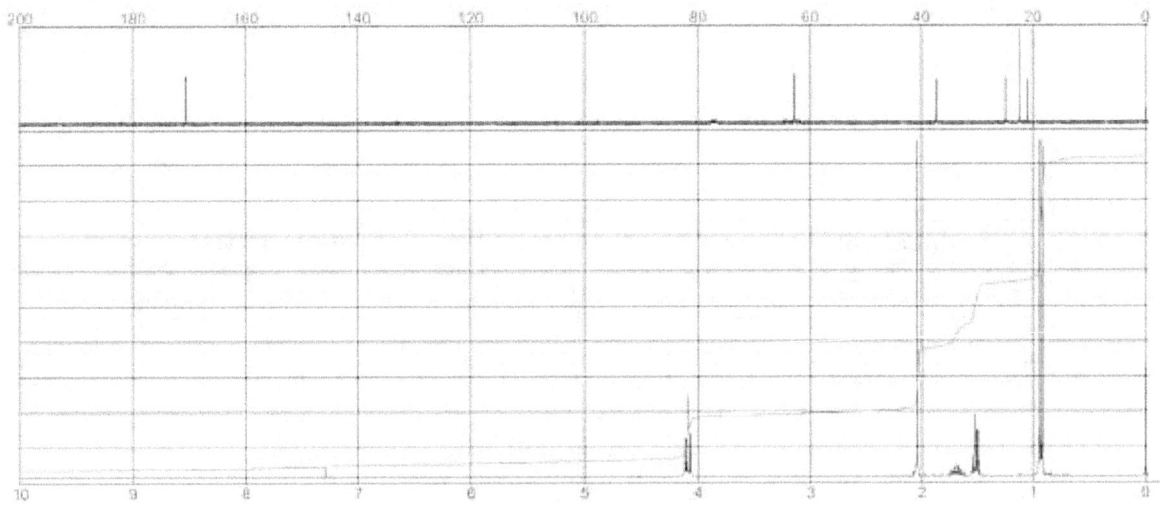

^{13}C NMR Spectrum

We can assign the peaks as follows:

Chemical shift δ (ppm)	Integral	Assignment(s)
171	1	C = O
64	1	C – O
37	1	C – C
25	1	C – C
22	2	C – C
21	1	C – C

There are several things to note here:

- The presence of both a C = O and a C – O functional group clearly demonstrates that the compound is an ester;
- The peak of integral two at δ 22 ppm indicates the presence of a pair of chemically and magnetically equivalent methyl, – CH₃, groups. There cannot be any larger pairs of chemically or magnetically equivalent groups as there would also, clearly be more than one pair of signals of integral two due to the – CH₂ – CH₃ groups.
- Equally, importantly there are insufficient carbon atoms for such a structure.

This means that we can propose the following structure:

If we label the molecule above we can assign the peaks as follows:

Chemical shift δ (ppm)	Integral	Assignment(s)
171	1	C(b)
64	1	C(c)
37	1	C(a), C(d) or C(e)
25	1	C(a), C(d) or C(e)
22	2	C(f)
21	1	C(a), C(d) or C(e)

We cannot distinguish between the signals from the proposed C(a), C(d) or C(e) but we can use these signals to predict the ^{1}H nmr spectrum of the molecule.

^{1}H NMR Spectrum

Using new labelling, we can predict the following multiplets and integrals as tabulated below:

Chemical shift δ (ppm)	Integral	Multiplicity	Assignment(s)
$0.5 - 2$	3	Singlet	H(g)
$3 - 4.2$	2	Triplet	H(h)
$0.5 - 2$	2	Doublet of triplets	H(i)
$0.5 - 2$	1	Triplet of septets	H(j)
$0.5 - 2$	6	Doublet	H(k) and H(k')

The actual observed, expanded, ^{1}H nmr spectrum is displayed next followed by a consideration of the multiplets which deserve further explanation.

Expanded ^{1}H nmr spectrum

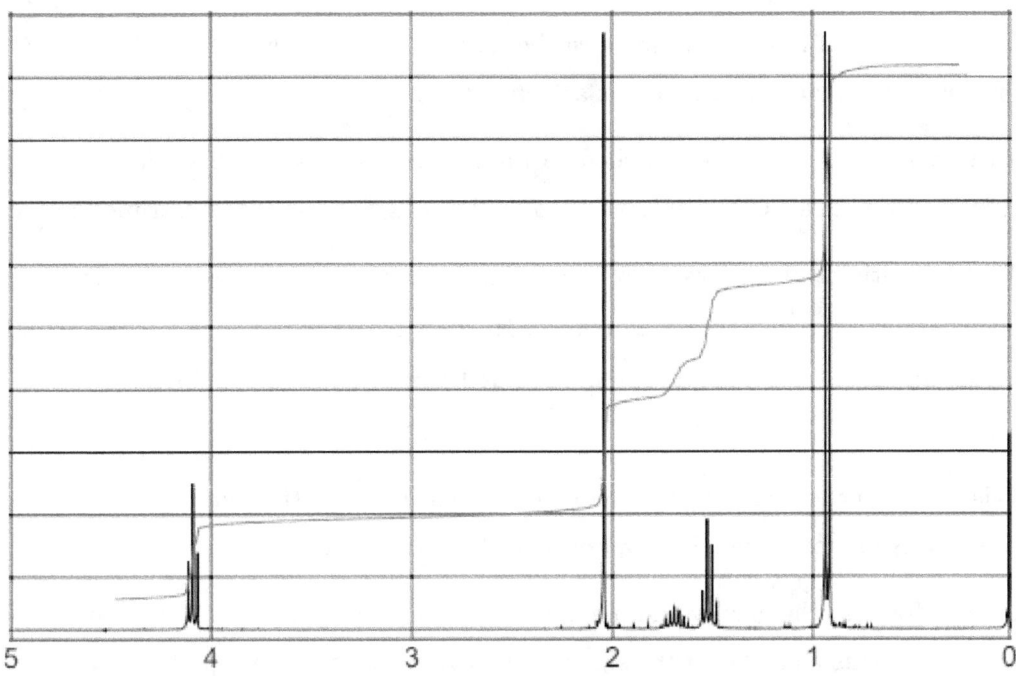

We can assign the peaks as follows:

Chemical shift δ (ppm)	Integral	Multiplicity	Assignment(s)
0.90	6	Doublet	H(k) and H(k')
1.50	2	Doublet of triplets	H(i)
1.65	1	Triplet of septets	H(j)
2.05	3	Singlet	H(g)
4.15	2	Triplet	H(h)

If we examine the region δ 1 – 2 ppm in more detail, shown below, we can see that the spectrum fits the predictions but we must also consider the multiplicities next.

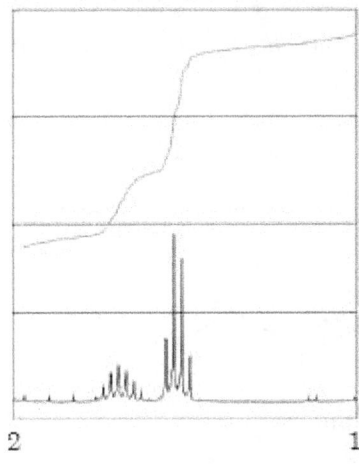

Application of the n+1 rule to explain the multiplicities:

- **H(g)** is a singlet of integral three since the methyl, $- CH_3$, group is attached to a $C = O$ carbon atom with no hydrogen atoms attached to it.

- **H(h)** is a triplet since it is attached to an oxygen atom, which has no effect on the multiplicity, and to a $- CH_2 -$ group and so the H(h) signal is split into a triplet (n+1 rule).

- **H(i)** is a multiplet of integral two because it is actually two hydrogen atoms and the bonded carbon atom is attached to two $- CH_2$ groups e.g. $H_2C - CH_2 - CH_2$. Using the labelling above H(i) is split into a triplet by H(h) and this is then split into a doublet of triplets by H(j).

- **H(j)** is bonded a carbon atom with hydrogen atoms on three other adjacent carbon atoms. These hydrogen atoms are labelled, above, as H(i), H(k) and H(k').

  H(k) and H(k') are chemically and magnetically equivalent and each label represents a $- CH_3$ group. This means that the signal due to J(j) is split into a septet by the combined total of six chemically methyl hydrogen atoms.

  This septet is split into a triplet of septets by the two hydrogen atoms labelled as H(i).

- **H(k)** and **H(k')** can be considered together as they are chemically and magnetically equivalent. It is important to remember that the $- CH_3$ group is tetrahedral and resembles a tripod which rotates freely and this explains why all six hydrogen atoms are chemically and magnetically equivalent and why it will give rise to a signal of integral six. The methyl groups represented by H(k) and H(k') are bonded to the same carbon atom which is, itself, bonded to a $- CH -$ group. The presence of the single hydrogen atom on this carbon atom would produce a doublet.

  This completely explain the structure of the molecule and so we can finalise our conclusions.

Conclusions

Structure:

Systematic name: isopentyl ethanoate.

Chapter XI

Ethyl caproate

Ethyl caproate is a pheromone employed by a variety of creatures including fruit flies (pictured above), apple humblebees and oriental fruit moths (pictured below).

Sparingly soluble in water, this substance is a colourless liquid with a melting point of -67°C and a boiling point of 168°C and it has the following physical properties:

Elemental composition: C: 66.57%, H: 11.21%; O: 22.19%

Formula mass (M_r): 144.2 g mol^{-1}.

These means that the molecule has the following compositions:

 Empirical formula: C_4H_8O

 Molecular formula: $C_8H_{16}O_2$

Infrared Spectrum

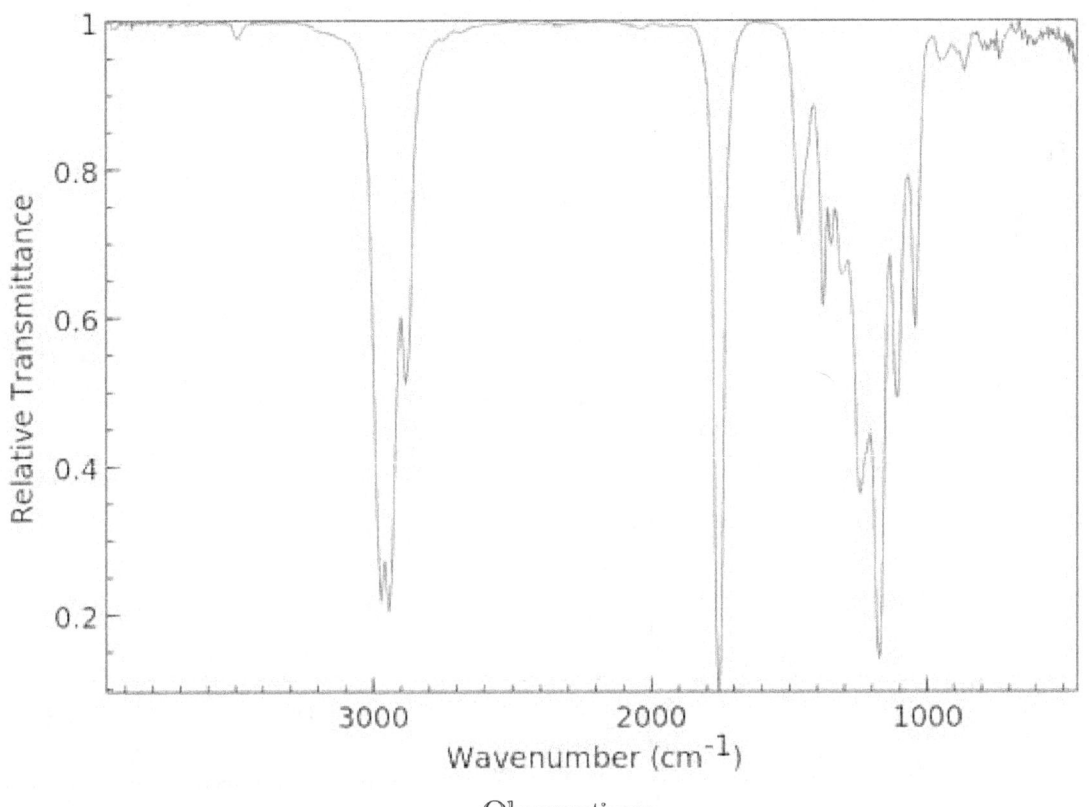

Observations

(√ /)	Wavenumber range (cm⁻¹)	Wavenumber (cm⁻¹)	Assignment
X	3200 - 3700		O – H
X	3200 - 3600		N – H
X	3000 – 3300		C – H (aromatic)
√	2500 – 3000	2980, 2905, 2820	C – H (aliphatic)
X	2200 – 2500		C ≡ N
√	1700 – 1800	1740	C = O
X	1600 – 1700		C = C (aliphatic)
X	1585 – 1600		C – C (aromatic)
X	1450 – 1600		C – C (aromatic)
√	1000 – 1300	1180	C – O
X	700 – 1000		C – X (X = Cl, Br or I)

Conclusions

Since the molecule contains two oxygen atoms and there are no aromatic C – H stretches
this molecule must be an aliphatic ester.

Mass Spectrum

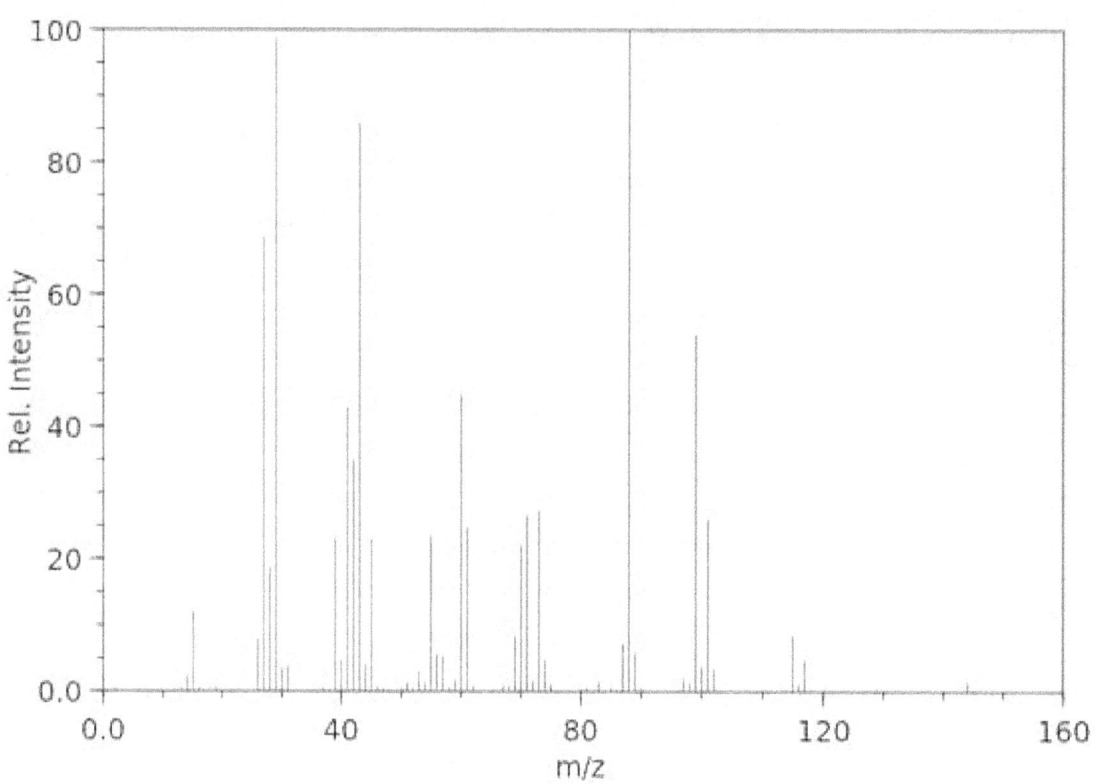

Observations

Charged fragments (m/z)	Assignment	Charged fragments (m/z)	Assignment
Molecular ion: 144	$[C_8H_{16}O_2]^+$	**Base peak: 88**	$[C_4H_8O_2]^+$
115	$[C_8H_3O]^+$	60	$[C_2H_4O_2]^+$
101	$[C_7HO]^+$	43	$[C_3H_7]^+$
99	$[C_7H_{13}]^+$	29	$[C_2H_5]^+$
71	$[C_5H_{11}]^+$	15	$[CH_3]^+$

Conclusions

The mass spectrum supports the suggestion from the infrared spectrum that the molecule is a long chain aliphatic ester but we can learn more from the ^{1}H and ^{13}C nmr spectra which we consider next.

Ethyl caproate

NMR Spectra

The ^{1}H and ^{13}C nmr spectra are shown below

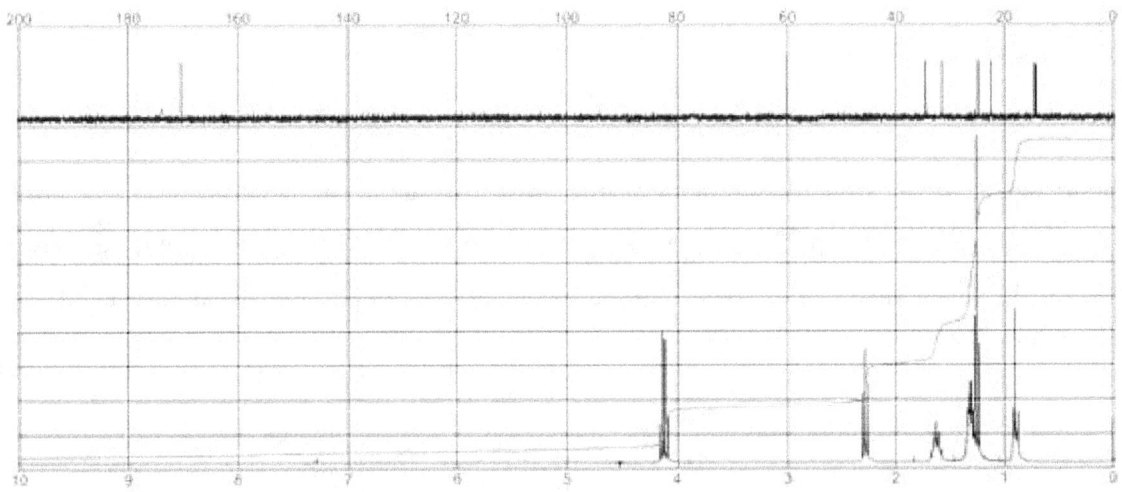

and we will examine the ^{13}C nmr spectrum first as we can readily assign some of the peaks.

Observations

^{13}C NMR Spectrum

Chemical shift δ (ppm)	Integral	Assignment(s)
171	1	C = O
60*	1	C – O
35	1	C – C
32	1	C – C
25	1	C – C
22	1	C – C
16	1	C – C
15	1	C – C

*This peak is exactly on the δ 60 ppm line of the spectrum's chart and so is not obviously apparent.

Conclusions

The peaks at δ 171 and δ 60 ppm demonstrate conclusively that the molecule is an ester and the number of aliphatic carbon signals suggests that it is an aliphatic long chain ester rather than a short chain, substituted molecule since all the C – C signals are of integral one.

Ethyl caproate

^{1}H NMR Spectrum

If we are correct that the molecule is an aliphatic long chain ester then the simplest candidate is:

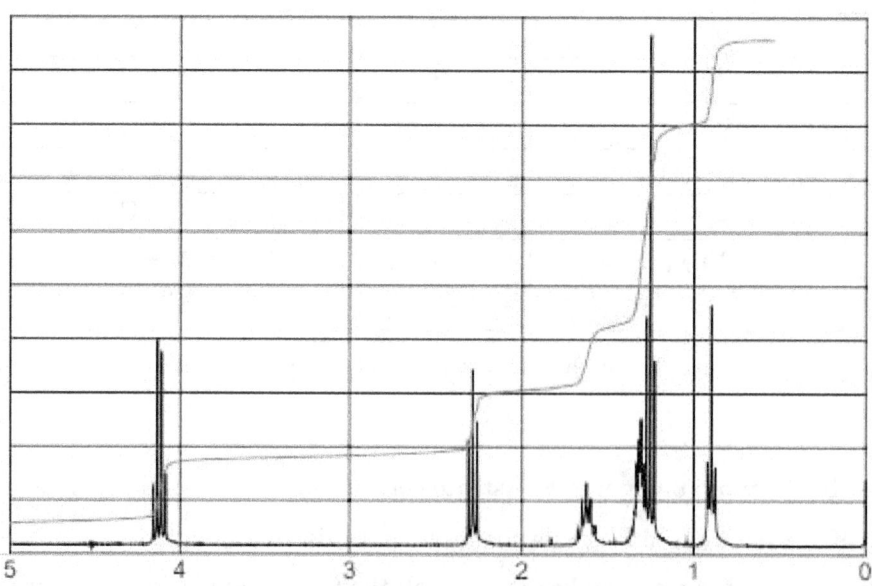

This means that there should a singlet of integral three due to the methoxy, $- OCH_3$, group in the region δ 3 – 4.2 ppm. There is no such signal and so this structure cannot be correct. There is, however, there is a quartet of integral two at δ 4.15 ppm and so the next candidate must be:

This reconciles nicely with the quartet at δ 4.15 ppm and there is also a triplet of integral three at δ 0.9 ppm and so together they account for an $- OCH_2CH_3$ group.

This leaves us with the task of analysing the rest of the spectrum which is repeated below and tabulated on the following page.

Observations and Conclusions

The ^{1}H nmr spectrum is tabulated below with the hydrogen atoms labelled alphabetically below.

Applying the n+1 rule we can make the following assignments:

Chemical shift δ (ppm)	Multiplet	Integral	Assignment
4.15	Quartet	2	H(f)
2.30	Triplet	2	H(e)
1.60	Quintet	2	H(c)
1.30	Overlapping Heptet / Quintet	4	H(b) / H(d)
1.25	Triplet	3	H(a)
0.90	Triplet	3	H(g)

since, working from the left hand side of the molecule as it is drawn above, we can conclude:

- **H(a)** is split into a triplet by H(b);
- **H(b) / H(d):**
    - **H(b)** is split into a heptet by H(a) and H(c) but we can also describe it as H(b) being split into a quartet by (Ha) with this quartet split into a triplet by H(c).
    - **H(d)** is split into a quintet by H(c) and H(e) but it can also be described as a triplet due to splitting by H(c) which is split into a triplet of triplets by H(e). The smallest peaks of the multiplet would not be visible unless the spectrum was expanded many more times than we can do here.

The chemical and magnetic environments of H(b) and H(d) are extremely similar and so the multiplets will overlap making it extremely difficult to separate out all of the peaks.

- **H(c)** will appear as a quintet since it is split by the chemically and magnetically equivalent H(b) and H(d).
- **H(e)** is a triplet because it is split by H(d)
- **H(f)** is a quartet due to splitting by H(g)
- **H(g)** is a triplet due to splitting by H(f).

Conclusions

Structure:

Systematic name: Ethyl hexanoate.

Chapter XII

Cetyl alcohol

Cetyl alcohol is a waxy white solid whose name derives from the Latin for whale oil from which it was first extracted and isolated in 1817 by Michel Chevreul from sperm whale(pictured above) oil. Industrially, it is now produced from an extract of palm oil.

Cetyl alcohol is widely used in the cosmetic and personal hygiene industries as an emulsifier and thickening agent. It is also used in some treatments for eczema.

It has been detected as yet another example of a queen bee retinue pheromone which encourages other bees to circle the queen forming a *retinue* and is also a precursor to *bombyol*, the first pheromone to have been isolated, identified and characterised by by Adolf Butenandt in 1959.

This compound has melting and boiling points of 49°C and 344°C respectively and the following properties:

Elemental composition: C: 79.21%, H: 14.17%; O: 6.61%
Formula mass (M_r): 242.4 g mol^{-1}.

This means that the empirical and molecular formulas are both $C_{16}H_{34}O$.

The presence of a large number of carbon atoms implies the existence of a long, or shorter substituted, chain or an aromatic compound whilst the presence of a single oxygen atom suggests that it is an ether or an alcohol. It cannot be a carboxylic acid or an ester since that would require two oxygen atoms in the molecular formula.

Cetyl alcohol

Infrared Spectrum

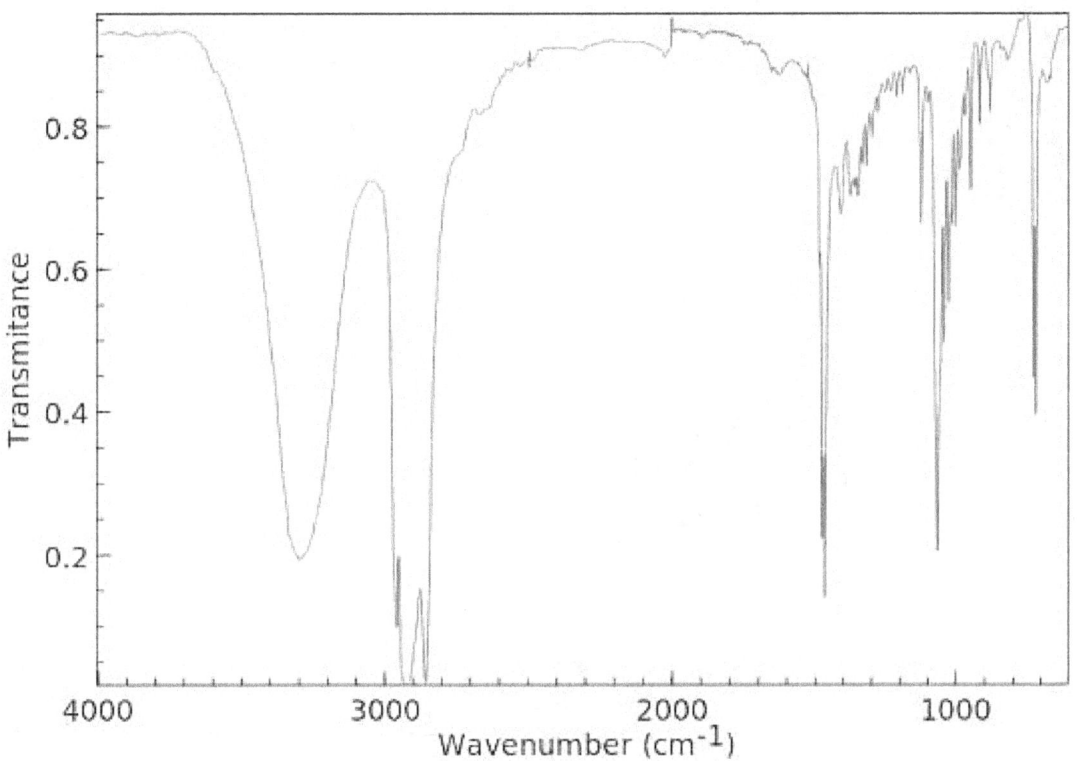

Observations

(√ / X)	Wavenumber range (cm⁻¹)	Wavenumber (cm⁻¹)	Assignment
√	3200 - 3700	3300	O – H
X	3200 - 3600		N – H
X	3000 – 3300		C – H (aromatic)
√	2500 – 3000	2980, 2920, 2850	C – H (aliphatic)
X	2200 – 2500		C ≡ N
X	1700 – 1800		C = O
X	1600 – 1700		C = C (aliphatic)
X	1585 – 1600		C – C (aromatic)
X	1450 – 1600		C – C (aromatic)
√	1000 – 1300	1050	C – O
X	700 – 1000		C – X (X = Cl, Br or I)

Conclusions

This molecule is clearly an aliphatic alcohol, a conclusions which is further confirmed by the presence of the peak at 1450 cm⁻¹ which is assignable to a C – O bond.

Cetyl alcohol

Mass Spectrum

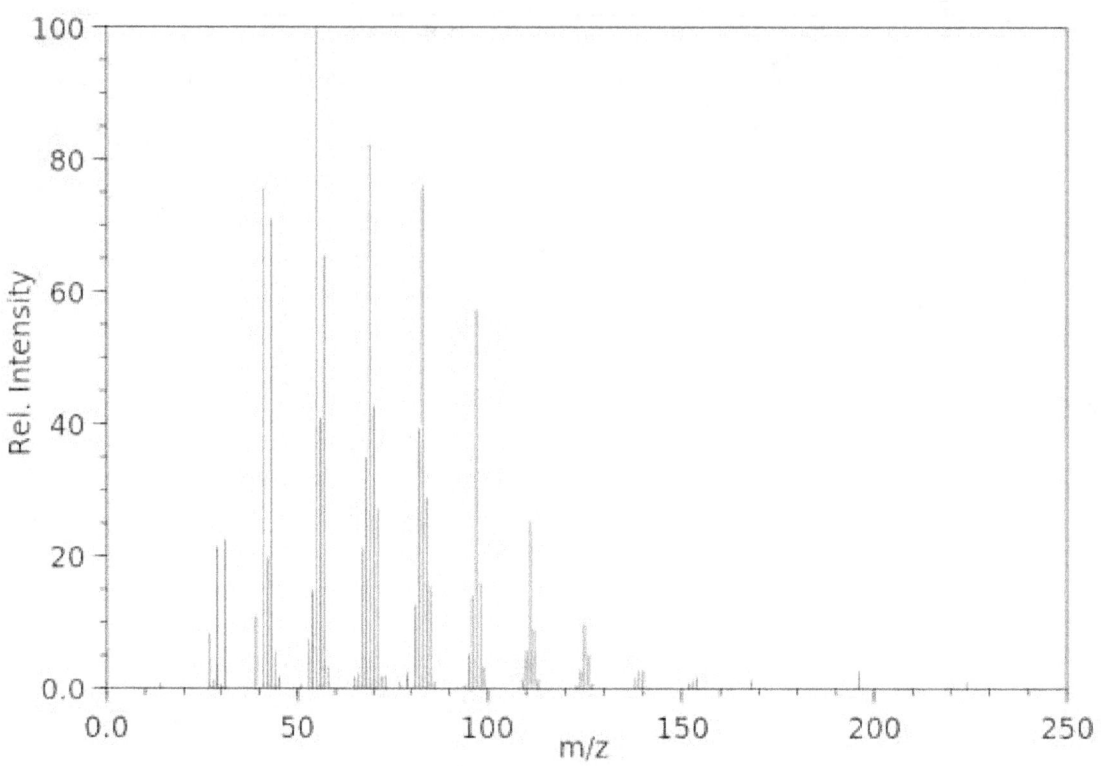

Observations

Charged fragments (m/z)	Assignment	Charged fragments (m/z)	Assignment
Molecular ion: 224	$[C_{16}H_{34}O]^+$	**Base peak: 55**	$[C_4H_7]^+$
158	$[C_{11}H_{26}]^+$	83	$[C_6H_{11}]^+$
154	$[C_{11}H_{22}]^+$	69	$[C_5H_9]^+$
140	$[C_{10}H_{20}]^+$	57	$[C_4H_9]^+$
125	$[C_9H_{17}]^+$	43	$[C_3H_7]^+$
111	$[C_8H_{15}]^+$	39	$[C_3H_3]^+$
97	$[C_7H_{13}]^+$	29	$[C_2H_5]^+$

Conclusions

From the high m/z peaks, the mass spectrum demonstrates that this molecule has a long, aliphatic, chain but we can learn more from the ¹H and ¹³C nmr spectra and, after having an overview of both spectra, we will start with the ¹³C nmr spectrum.

NMR Spectra

The ^1H and ^{13}C nmr spectra are displayed below:

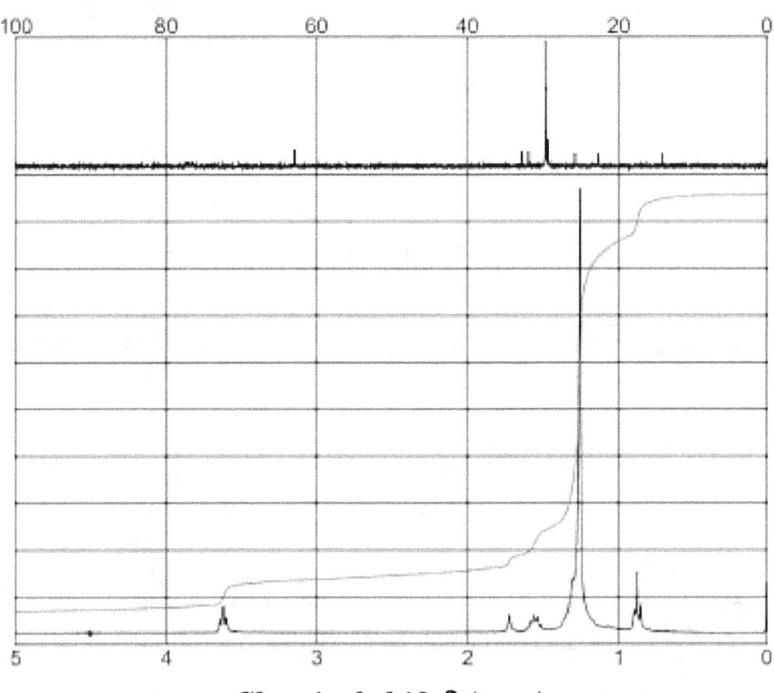

Chemical shift δ (ppm)

We will examine the ^{13}C nmr spectrum first.

^{13}C NMR Spectrum

Examining the expanded ^{13}C nmr spectrum,

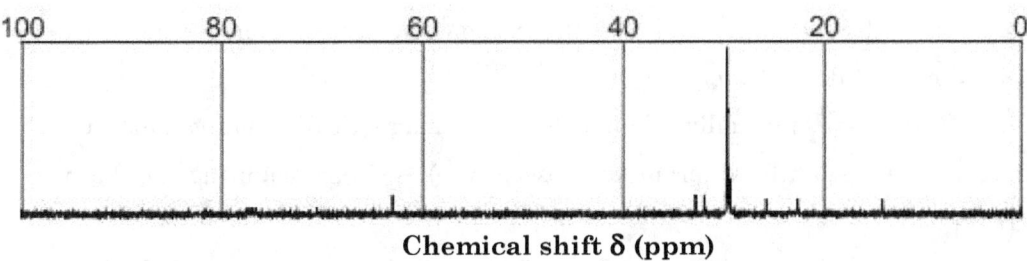

Chemical shift δ (ppm)

▪ The peak at δ 63 ppm is due to a C – O group which must be due to the alcoholic, – C – OH, group;

▪ The peak at δ 15 ppm is due to an alkyl group which may be at the other end of the molecule;

The remaining signals are also in the alkyl, – C – C – , region and the peak at δ 30 ppm is particularly interesting since it is of integral eight implying there are eight – CH_2 – groups.

This suggests that the molecule has the following structure:

We can assign the peaks as follows, using the same structure but now labelled alphabetically:

Chemical shift δ (ppm)	Integral	Assignment(s)
63	1	C(a)
34	1	C(b)
33	1	C(c)
30	8	C(e) – C(l)
29	2	C(d) / C(m)
25	1	C(n)
22	1	C(o)
15	1	C(p)

The assignments can be rationalised as follows:

- The highlighted rectangle identifies eight chemically and magnetically equivalent carbon atoms, C(e) – C(l);

- C(a) has already been assigned and the, electronegative, oxygen atom deshields C(a) and, to a lesser extent C(b) and C(c);

- C(d) and C(m) are mutually chemically and magnetically equivalent but their environments are slightly different to those of C(e) – C(l), accounting for the peak of integral of two;

- C(p) is the most alkyl carbon atom and so will have the lowest chemical shift;

- The remaining alkyl signals can be assigned to C(n) and C(o).

The next task is to determine if the conclusions are supported by the ^{1}H nmr spectrum.

We will examine the structure to predict the spectrum before examining the actual, observed spectrum.

Cetyl alcohol

^{1}H NMR Spectrum

If the proposed structure is correct then we can make the following predictions using capital letters to identify the hydrogen atoms:

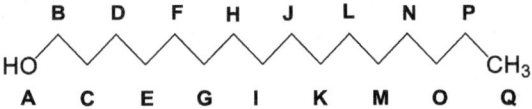

Chemical shift δ (ppm)	Multiplicity	Integral	Assignment
0 – 12	Singlet	1	H(A)*
3 – 4.2	Triplet due to splitting by H(C)	2	H(B)
0.5 – 2	Triplet of triplets due to splitting by H(C) and H(D)	2	H(C)
0.5 – 2	Triplet of triplets due to splitting by H(C) and H(E)	2	H(D)
0.5 – 2	Triplet of triplets due to splitting by H(D) and H(F)	2	H(E)
0.5 – 2	Triplet of triplets due to splitting by H(E) and H(G)	2	H(F)
0.5 – 2	Triplet of triplets due to splitting by H(F) and H(H)	2	H(G)
0.5 – 2	Triplet of triplets due to splitting by H(G) and H(I)	2	H(H)
0.5 – 2	Triplet of triplets due to splitting by H(H) and H(J)	2	H(I)
0.5 – 2	Triplet of triplets due to splitting by H(I) and H(K)	2	H(J)
0.5 – 2	Triplet of triplets due to splitting by H(J) and H(L)	2	H(K)
0.5 – 2	Triplet of triplets due to splitting by H(K) and H(M)	2	H(L)
0.5 – 2	Triplet of triplets due to splitting by H(C) and H(D)	2	H(M)
0.5 – 2	Triplet of triplets due to splitting by H(M) and H(O)	2	H(N)
0.5 – 2	Triplet of triplets due to splitting by H(N) and H(P)	2	H(O)
0.5 – 2	Triplet of quartet due to splitting by H(C) and H(D)	2	H(P)
0.5 – 2	Triplet due to splitting by H(P)	3	H(Q)

*Hydroxyl, – OH, hydrogen atoms often do not appear in ^{1}H nmr spectra due to rapid exchange with the solvent hydrogen atoms and when they do appear they can appear anywhere in the spectrum.

Cetyl alcohol

If we examine the actual, expanded, ^{1}H nmr spectrum we see that the structure is confirmed:

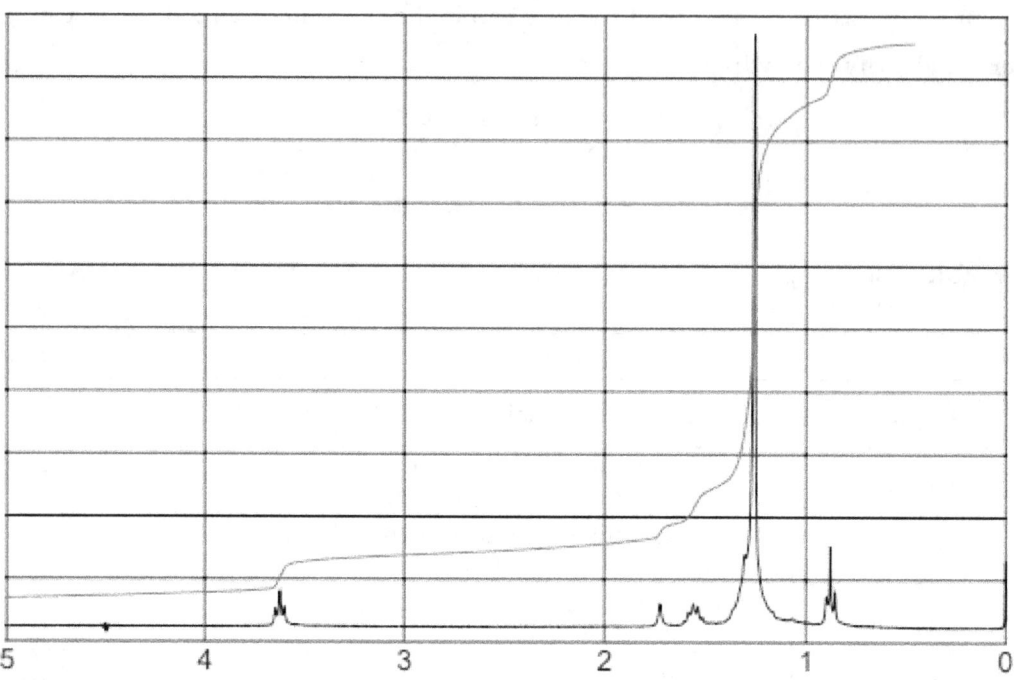

as discussed next.

Chemical shift δ (ppm)	Multiplicity	Integral	Assignment
3.65	Triplet due to splitting by H(C)	2	H(B)
1.75	Singlet	1	H(A)
1.60	Sextet	2	H(P)
1.25	Triplet of triplets due to splitting by H(C) and H(D)	26	H(C) – H(O)
0.90	Triplet due to splitting by H(P)	3	H(Q)

This is all consistent with the proposed structure.

Conclusions

Structure:

HO⌇⌇⌇⌇⌇⌇⌇CH₃

Systematic name: Hexadecan-1-ol

Chapter XIII

Nerol

This substance is a cockroach (picturedabove) pheromonewhich is secreted by females to attract mates. It is also very effective as a pheromone of the stingless bee, *Trigona fulviventris* (pictured below), when it works in conjunction with octyl caproate.

With a melting point of -15°C and a boiling point of 314°C, nerol is a colourless liquid with the following properties:

Elemental composition: C: 77.80%, H: 11.79%; O: 10.37%

Formula mass (M_r): 154.25 g mol^{-1}.

This means the empirical formula and molecular formulas are both $C_{10}H_{18}O$.

The number of carbon atoms indicates an aliphatic long, or shorter substituted, alkyl chain or an aromatic compound. The presence of only one oxygen atom means that the molecule must be an ether or an alcohol. It cannot be an ester or a carboxylic acid as that structure would require two oxygen atoms in the molecular formula.

Nerol

Infrared Spectrum

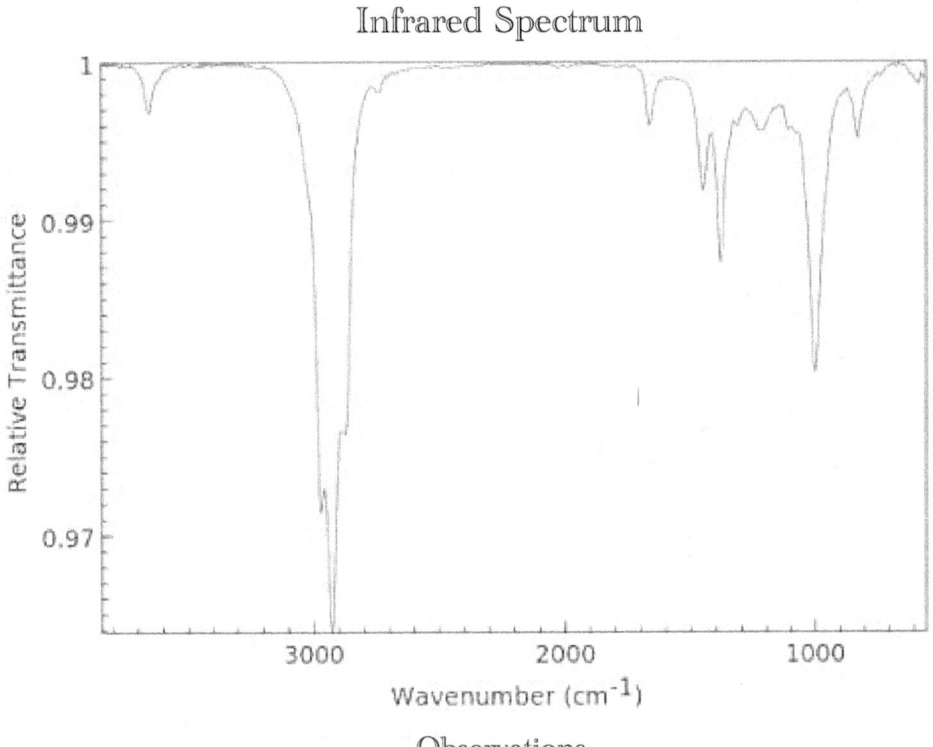

Observations

(√ / X)	Wavenumber range (cm⁻¹)	Wavenumber (cm⁻¹)	Assignment
√	3200 - 3700	3620	O – H
X	3200 - 3600		N – H
X	3000 – 3300		C – H (aromatic)
√	2500 – 3000	2990, 2925	C – H (aliphatic)
X	2200 – 2500		C ≡ N
X	1700 – 1800		C = O
X	1600 – 1700	1630	C = C (aliphatic)
X	1585 – 1600		C – C (aromatic)
X	1450 – 1600		C – C (aromatic)
√	1000 – 1300	1005	C – O
X	700 – 1000		C – X (X = Cl, Br or I)

Conclusions

The infrared spectrum clearly suggests that this molecule is an aliphatic alcohol. It cannot be an ester or carboxylic acid as there is only one oxygen atom per molecule.

The peak at 3620 cm⁻¹ may be due to the presence of an – OH group but it is unusually weak as is the C = C peak at 1630 cm⁻¹. This means that the aliphatic alcohol chain is unsaturated.

Nerol

Mass Spectrum

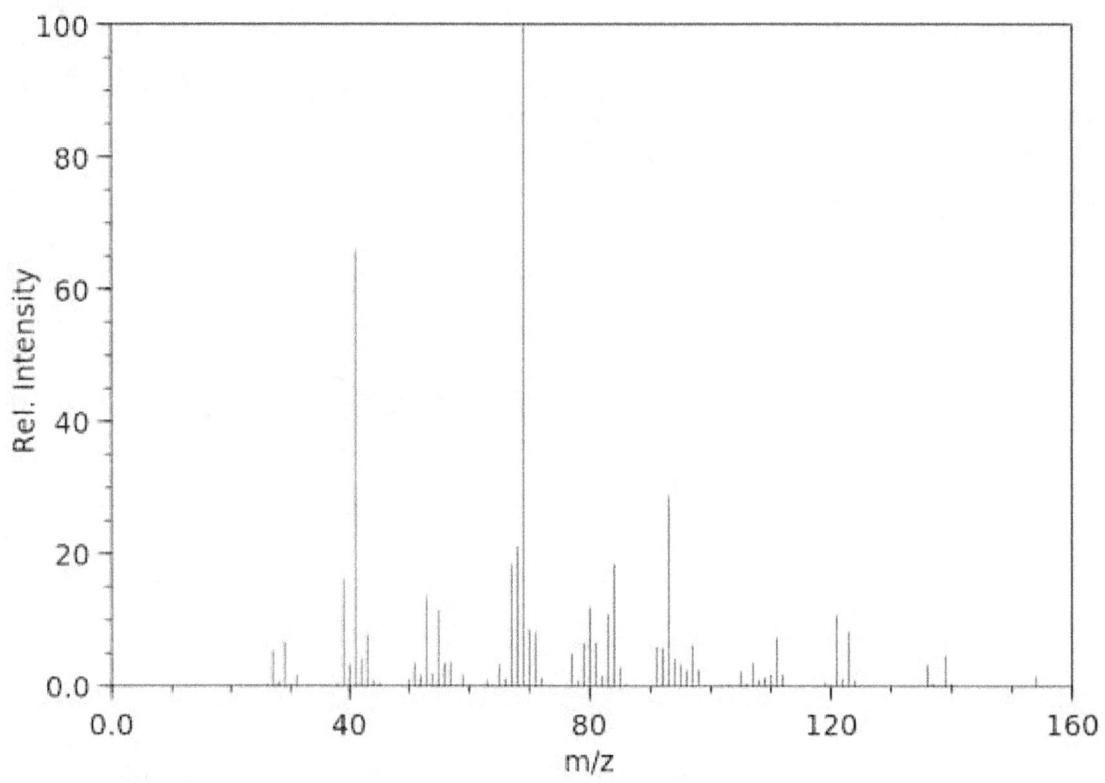

Observations

Charged fragments (m/z)	Assignment	Charged fragments (m/z)	Assignment
Molecular ion: 154	$[C_{10}H_{18}O]^+$	**Base peak: 69**	$[C_5H_9]^+$
139	$[C_9H_{15}O]^+$	84	$[C_6H_{12}]^+$
123	$[C_9H_{15}]^+$	55	$[C_4H_7]^+$
111	$[C_8H_{15}]^+$	41	$[C_3H_5]^+$
93	$[C_7H_9]^+$	39	$[C_3H_3]^+$

Conclusions

The mass spectrum supports the supposition that this molecule is a long chain aliphatic alcohol but adds little to the determination of the structure.

We can learn more from the ^1H and ^{13}C nmr spectra but we can postulate a structure before that task.

NMR Spectra

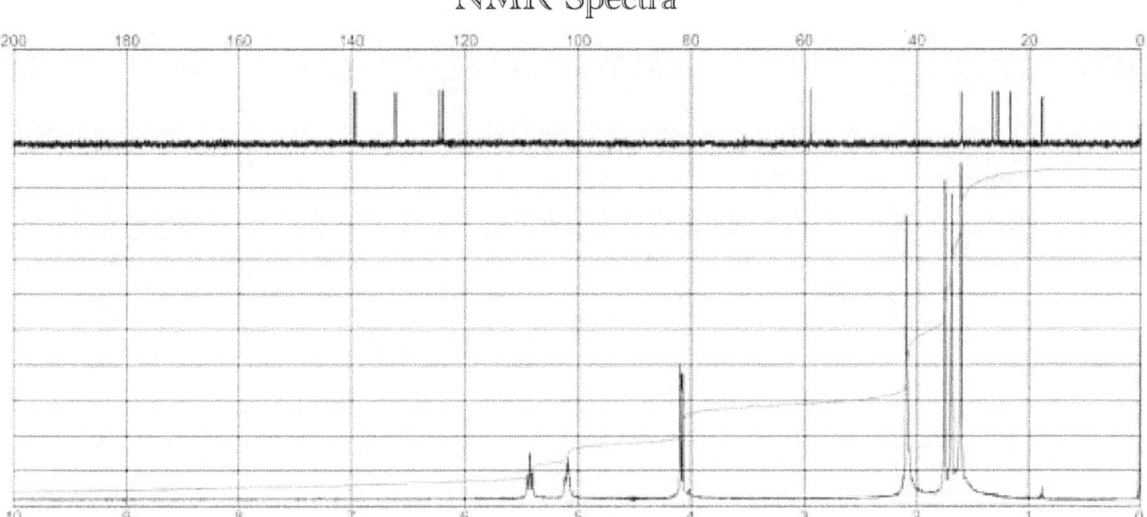

The simplest possible structure for this molecule comprises a ten carbon – aliphatic chain terminating in an – OH and with at least one C = C bond. There are too many possibilities to predict a structure so it is better to examine the nmr spectra first and we will start with the ^{13}C nmr spectrum.

^{13}C NMR Spectrum

There are ten signals each of integral one which confirms the Mr of the compound but we can make the following conclusions. There:

- Are four C = C carbon atoms in the region δ 110 – 160 ppm, indicating the presence of two C = C bonds;
- Is one C – OH carbon atom, δ 58 ppm, confirming that the molecule is an alcohol;
- Are five alkyl carbon atoms which cause the signals in the δ 0 – 50 ppm region.

If we now start with the, justifiable, assumption that the molecule is a long chain, unsaturated alcohol comprising ten carbon atoms but with two C = C bonds there are at least thirteen possible candidate molecules.

If the assumption is correct then the backbone will be

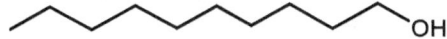

but with the replacement of two C – C bonds with two C = C bonds. It is clearly a waste of time to predict the spectra of all thirteen isomers and it is much more efficient to examine the ^{1}H nmr spectrum which is our next task.

^{1}H NMR Spectrum

If we examine the spectrum in expanded detail we observe the following from the C = C and C − O region:

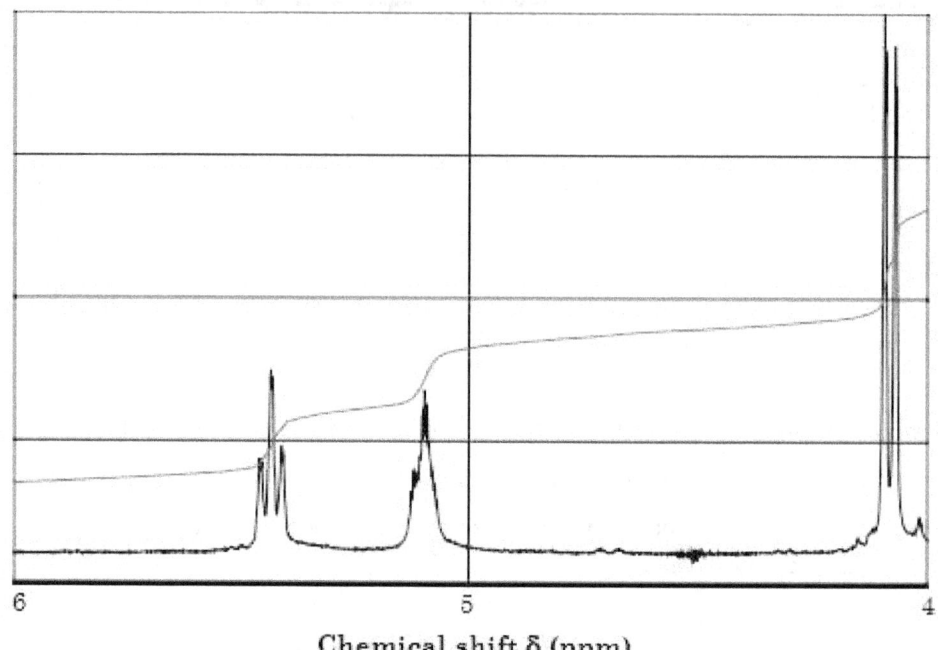

Chemical shift δ (ppm)

- The multiplet at δ 4.1 ppm is a doublet of integral two and so the terminal hydroxyl group must be bonded to a − CH$_2$ − group i.e.

$$HO - CH_2 -$$

- There are two chemically and magnetically distinct alkene hydrogen atoms:-
    - One forms a triplet of integral one which indicates that this hydrogen atom only has two hydrogen atoms on the two adjacent carbon atoms. This indicates that part of the structure could be

$$HO - CH_2 - CH = C <$$

It is better to draw this fragment as:

$$HO - CH_2 - CH = C \overset{|}{-}$$

as this indicates that one of the, so far, undetermined groups must continue the chain and the other must be a terminal − CH$_3$ group.

We now have a plausible structural fragment and can consider the other C = CH signal which is a triplet. This means that the other C= C bond cannot be immediately adjacent and so there must be a – CH₂ – group on the right hand side of the molecule as shown below:

$$HO - CH_2 - CH = C - CH_2 -$$
$$|$$

This now accounts for five of the carbon atoms and nine of the hydrogen atoms and so it is worth looking at the alkyl region to establish exactly where these hydrogens can lie in the molecule.

The expanded region is shown below.

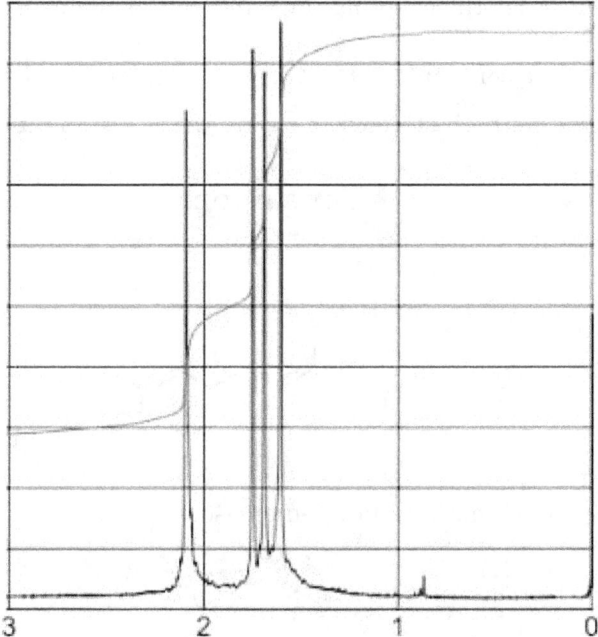

There are four signals of integrals 3:2:2:3.

The two signals of integral three are singlets so they can only be caused by two methyl, – CH₃ groups attached to a carbon atom with no hydrogen atoms on that atom.

This can arise from a terminal carbon atom with two methyl substituents and no hydrogen atoms on that carbon atom.

This must mean that we have the following terminal group:

$$= C - (CH_3)_2$$

This leaves us with two triplets of $C - CH_2$ groups which must fit in between the two groups we have established and so the only feasible structure is as shown below:

This also permits us to assign the second alkene hydrogen atom which is a triplet and is the alkene hydrogen atom on the right hand side of the molecule as drawn.

Coupling constants

The last matter to address is the coupling constants of the alkene hydrogen signals.

In both cases, the coupling constants of the alkene multiplets are $J = 7$ Hz and, from the data sheet, we can conclude that both stereoisomers are the Z – configuration.

Conclusions

Structure:

Systematic name: (Z)-3,7-dimethylocta-2,6-dien-1-ol

Chapter XIV

Ethyl oleate

Ethyl oleate is a fascinating ester which is an important method of communication employed by honey bees.

This hydrophobic compound, a light yellow oil with a slight floral fragrance has a melting point of -32°C and a boiling point of 216°C.

Elemental composition: C: 77.29%, H: 12.36%; O:10.31%

Formula mass (M_r): 310.5 g mol^{-1}.

These means that it has the following properties:

Empirical formula: $C_{10}H_{19}O$

Molecular formula: $C_{20}H_{38}O_2$.

All esters are produced by reaction of an alcohol and an acid and bees produce it from the reaction between ethanol, produced from flowers from the breakdown of sugars, and fatty acids. The human body also produces this molecule after ingestion of alcohol and it has been suggested that it is one mediator leading to foetal alcohol syndrome.

Ethyl oleate

Infrared Spectrum

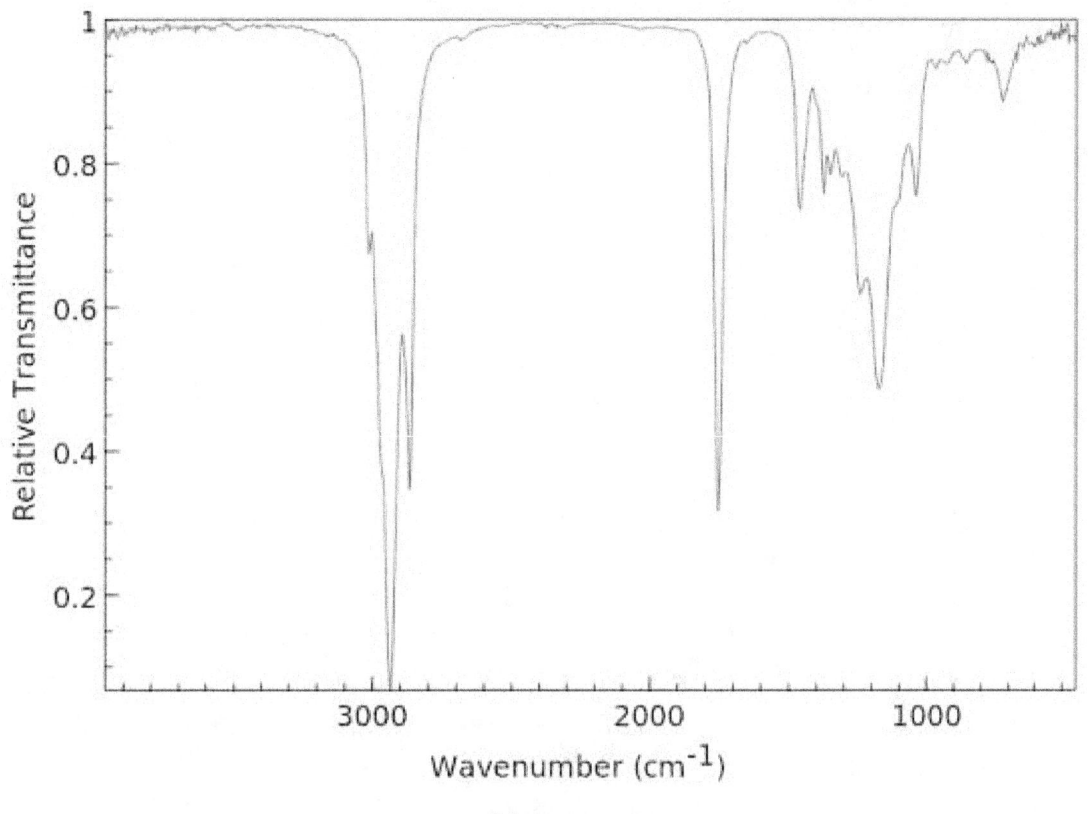

Observations

(√ / X)	Wavenumber range (cm⁻¹)	Wavenumber (cm⁻¹)	Assignment
X	3200 - 3700		O – H
X	3200 - 3600		N – H
X	3000 – 3300		C – H (aromatic)
√	2500 – 3000	2920, 2860	C – H (aliphatic)
X	2200 – 2500		C ≡ N
√	1700 – 1800	1750	C = O
X	1600 – 1700		C = C (aliphatic)
X	1585 – 1600		C – C (aromatic)
X	1450 – 1600		C – C (aromatic)
√	1000 – 1300	1170	C – O
X	700 – 1000		C – X (X = Cl, Br or I)

Conclusions

The presence of C – H, C = O and C – O peaks indicate this molecule is an aliphatic ester

which is supported by the presence of two oxygen atoms in the molecular formula.

Ethyl oleate

Mass Spectrum

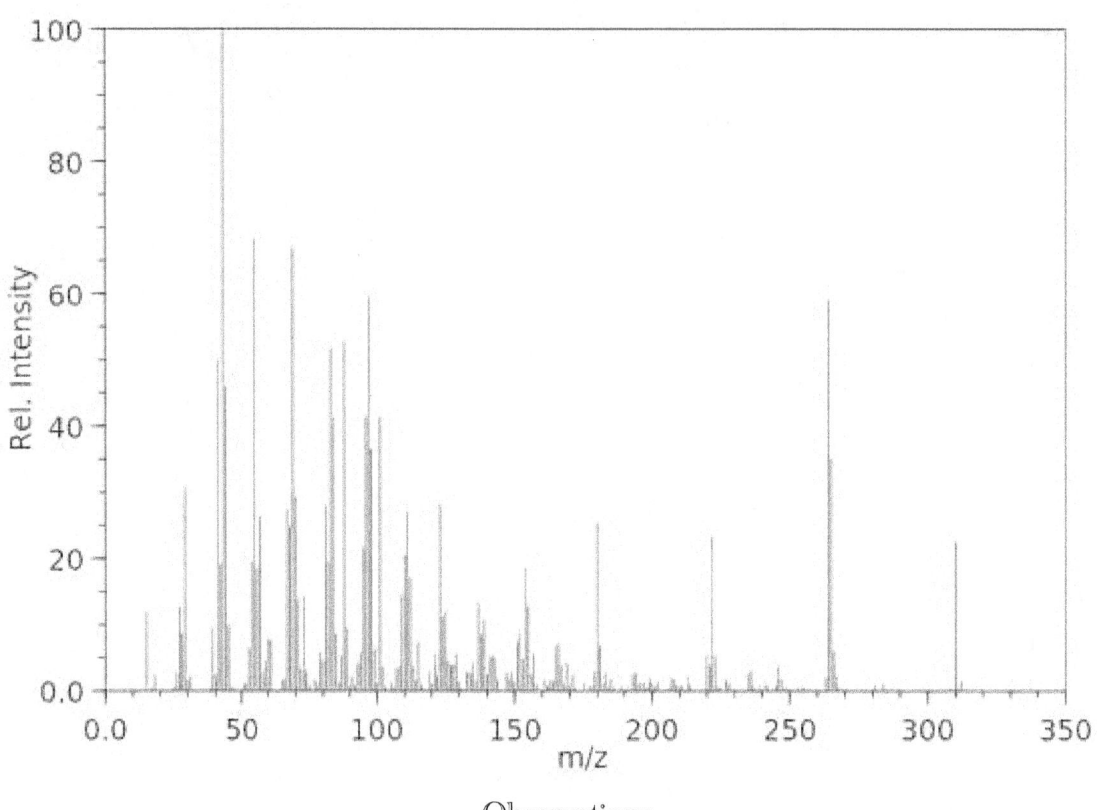

Observations

Charged fragments (m/z)	Assignment	Charged fragments (m/z)	Assignment
Molecular ion: 310	$[C_{20}H_{38}O_2]^+$	**Base peak: 43**	$[C_{20}H_{38}O_2]^+$
264	$[C_{17}H_{28}O_2]^+$	97	$[C_7H_{13}]^+$
222	$[C_{16}H_{30}]^+$	88	$[C_4H_9O_2]^+$
180	$[C_{11}H_{16}O_2]^+$	83	$[C_6H_9]^+$
154	$[C_{10}H_{18}O]^+$	69	$[C_5H_9]^+$
111	$[C_8H_{15}]^+$	55	$[C_4H_7]^+$
101	$[C_7H_9]^+$	29	$[C_2H_5]^+$

Conclusions

The occurrence of many fragments of varying numbers of oxygen atoms suggests that this molecule is a long chain ester. An ester can have chains of any length on both sides of the functional group and so the nmr spectra are essential to determine the structure. We examine the spectra next.

Ethyl oleate

NMR Spectra

We will consider the ¹H nmr spectrum first.

¹H NMR Spectrum

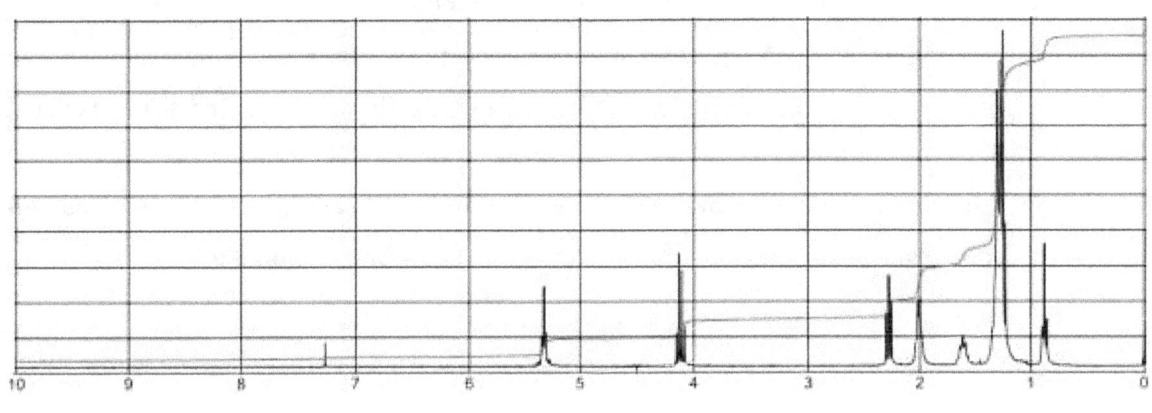

Observations

- Firstly, we can disregard the small peak at δ 7.3 ppm which is due to a contamination in the deuterated chloroform solvent which is very common.

- The peak at δ 5.4 ppm, of integral two, is clearly due to two alkene hydrogen atoms and is a doublet of triplets with coupling constants of J = 7 and 11 Hz.

  - This indicates that the fragment is Z – (zusammen) and that the fragment has the structure:

$$\sim\!\!\sim\!\!\underset{H_2}{C}\!-\!\overset{H}{C}\!=\!\overset{H}{C}\!-\!\underset{H_2}{C}\!\!\sim\!\!\sim$$

where the squiggle denotes an undefined group.

We can explain the doublet of triplets since each of the alkene hydrogen atoms, is split into a triplet by the adjacent CH_2 – group. In other words, using the diagram below, H(a) is split into a triplet by the two H(c) atoms, (n+1 rule) and this triplet is split into a doublet of triplets by the other alkene hydrogen atom, H(b). The hydrogen atoms H(d) are too far away to have any impact.

$$H_2\underset{c}{C}\!-\!\overset{H_a}{C}\!=\!\overset{H_b}{C}\!-\!\underset{H_2\,d}{C}\!\!\sim\!\!\sim$$

The same applies if we work from H(b) which is split into a triplet by H(d) and then a doublet of triplets by H(a) and is unaffected by H(c).

Ethyl oleate

- The quartet, of integral two, at δ 4.15 ppm must be due to a – O – CH$_2$ – CH$_3$ group.
  - The quartet is caused by the – CH$_3$ group and the integral of two must be due to the – CH$_2$ – grouping.
- There are two triplets of integral three at δ 2.35 ppm and δ 0.92 ppm.
  - The triplet at δ 0.92 ppm can be assigned to a terminal – CH$_3$ group and so terminates the entire alkyl chain but, to explain the triplet, it *must* be attached to a – CH$_2$ group.
  - The triplet at δ 2.35 ppm must also be due to a – CH$_3$ group but is deshielded and so we can assign the following structure to the ester functional group:

This means that the overall structure can be generalised as:

$$CH_3 – CH_2 – C_{16}H_{28} – C(O) – O – C_2H_5$$

and visualised as:

where the squiggle, $\sim\sim\sim$ represents the chain of C$_{16}$H$_{30}$.

There is no evidence for a substituted chain but we can investigate the ^{1}H nmr spectrum for more clues.

- There is a triplet of triplets, of integral four at δ 2 ppm. Since it is of integral four then this must represent at least two chemically and magnetically hydrogen atoms bonded to carbon atoms with a total of three hydrogen atoms bonded to adjacent carbon atoms. This can only be explained by the following fragment:

and this means that we draw a rough structure as shown below:

Ethyl oleate

This structure is nice as it demonstrates how chemists mix and match different ways of drawing structures for greatest clarity. Again the squiggle indicates an undetermined structure and our last task is to determine the position of the C = C bond which can be in one of a number of places. Assuming that we are correct and that this molecule is a linear, long chain, ester, we have a number of possible positions for the C = C bond and hence a number of possible structures. Since the molecular formula is $C_{20}H_{38}O_2$, we now merely have seven carbon atoms to account for. To predict the spectra of all of these candidates would be extremely tedious, time – consuming and futile and it is better to look back at the ^{1}H and ^{13}C nmr spectra.

The multiplet centred on δ 1.25 ppm comprises overlying multiplets as follows:

* Four triplets of triplets, each of integral two;
* Four quintets, each of integral two.

The only possible structure to comply with these resonance signals is:

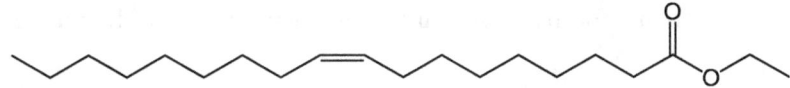

This structural assignment explains why there are so many chemically and magnetically equivalent hydrogen atoms and is confirmed by the ^{13}C nmr spectrum.

^{13}C NMR Spectrum

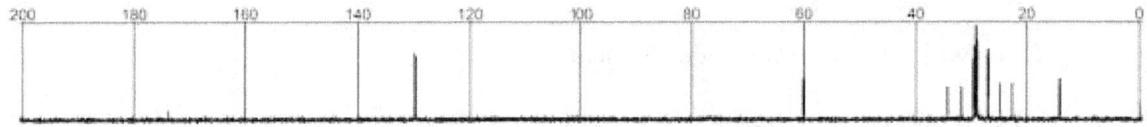

The proposed structure is supported by the ^{13}C nmr spectrum which shows the existence of:-

* The two C = C bond carbon atoms at δ 131 ppm which is a peak of integral two;
* The C = O carbon atom at δ 61 ppm of integral one;
* The collection of peaks of integral ten, at δ 31, 30 and 29 ppm can explained by the long chain skeleton. There is only structure in which six of the carbon atoms are chemically and magnetically equivalent. The remainder of the alkyl carbon atoms constitute the remainder of the peaks in the region δ 20 – 40 ppm with the exception of the terminal methyl, – CH3, group which appears at δ 15 ppm.

Coupling Constants

There is one more matter to consider and that is the orientation of the groups around the double bond. We can draw the molecule as either Z – (zusammen) or E – (entgegen) configurations as shown generalised below:

where R and R' are any group but cannot be hydrogen atoms.

The data sheet records that for alkenes in general the, experimentally measured, coupling constants, are:-

- Zusammen $(Z-)$: $J = 5 - 14$ Hz
- Entgegen $(E-)$: $J = 15 - 20$ Hz

The coupling constant of the alkene hydrogen in this molecule is $J = 7$ Hz and so this is the zusammen $(Z-)$ isomer.

Conclusions

Structure:

Systematic name: Ethyl (9Z)-octadec-9-enoate

Chapter XV

Geranyl acetate

Geranyl acetate is found in many creatures including yellow and black mud daubers (pictured above), cherry fruit flies, early mining bees, Indian bees Cullum's and early bumblebees as well in the chestnut gall wasp (pictured below) in which it acts as an alarm pheromone.

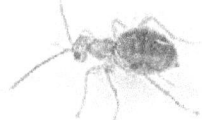

Alarm pheromones offer either a warning or announce a call for help and encourage species to marshal defences. Social insects, such as ants, bees, wasps and termites all demonstrated show colony defence behaviour which is coordinated by alarm pheromones when the colony is attacked or the colony's food supply is threatened. This was first realised with honey bees who work together to defend their honey from intruders.

With a light floral fragrance, geranyl acetate is also found in a wide range of plants including carrot, lemongrass and coriander.

Completely insoluble in water, this particular compound is a colourless liquid with a melting point of -100°C and a boiling point of 250°C and has the following properties:

Elemental composition: C: 73.36%, H: 10.29%; O: 16.30%
Formula mass (M_r) of 196.3 g mol^{-1}.

These means that it has the following formulas:

Empirical formula: $C_6H_{10}O$

Molecular formula: $C_{12}H_{20}O_2$.

Geranyl acetate

Infrared Spectrum

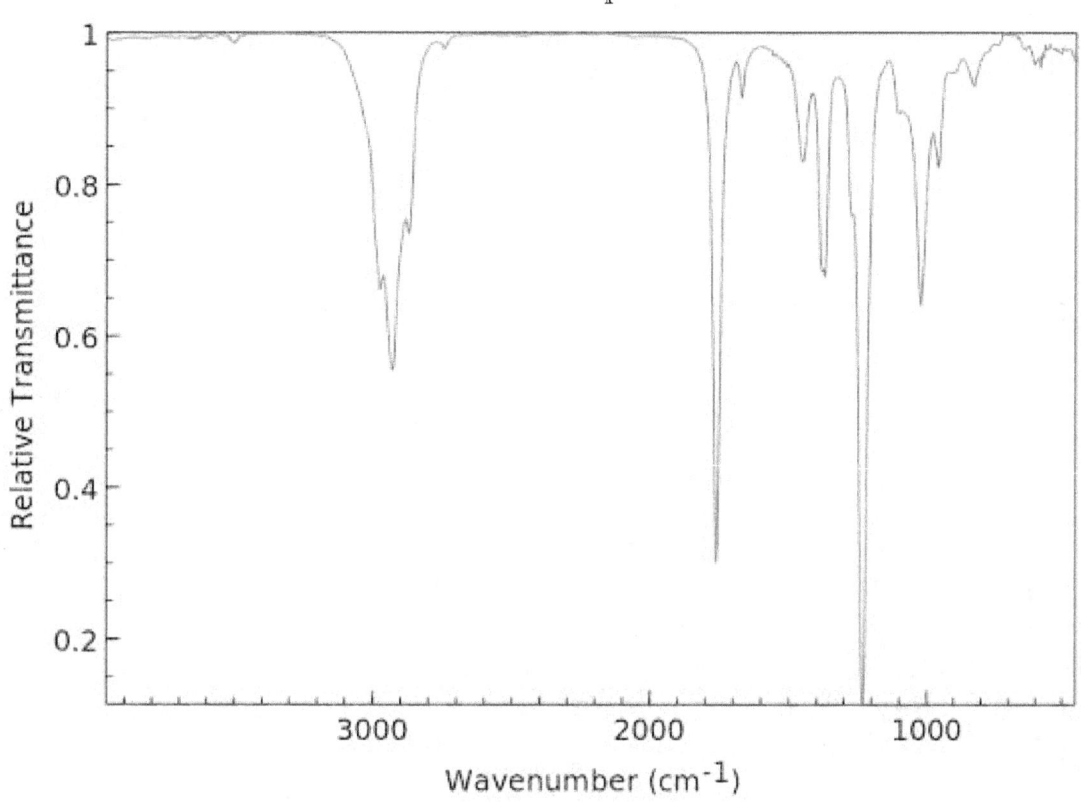

Observations

(√ / X)	Wavenumber range (cm⁻¹)	Wavenumber (cm⁻¹)	Assignment
X	3200 - 3700		O – H
X	3200 - 3600		N – H
X	3000 – 3300		C – H (aromatic)
√	2500 – 3000	3020, 2890, 2800	C – H (aliphatic)
X	2200 – 2500		C ≡ N
√	1700 – 1800	1750	C = O
√	1600 – 1700	1650	C = C (aliphatic)
X	1585 – 1600		C – C (aromatic)
X	1450 – 1600		C – C (aromatic)
√	1000 – 1300	1230	C – O
X	700 – 1000		C – X (X = Cl, Br or I)

Conclusions

We can determine that this molecule is an aliphatic ester with at least one C = C bond although the C = C signal is very weak, perhaps because of the strength of the C = O signal.

Geranyl acetate

Mass Spectrum

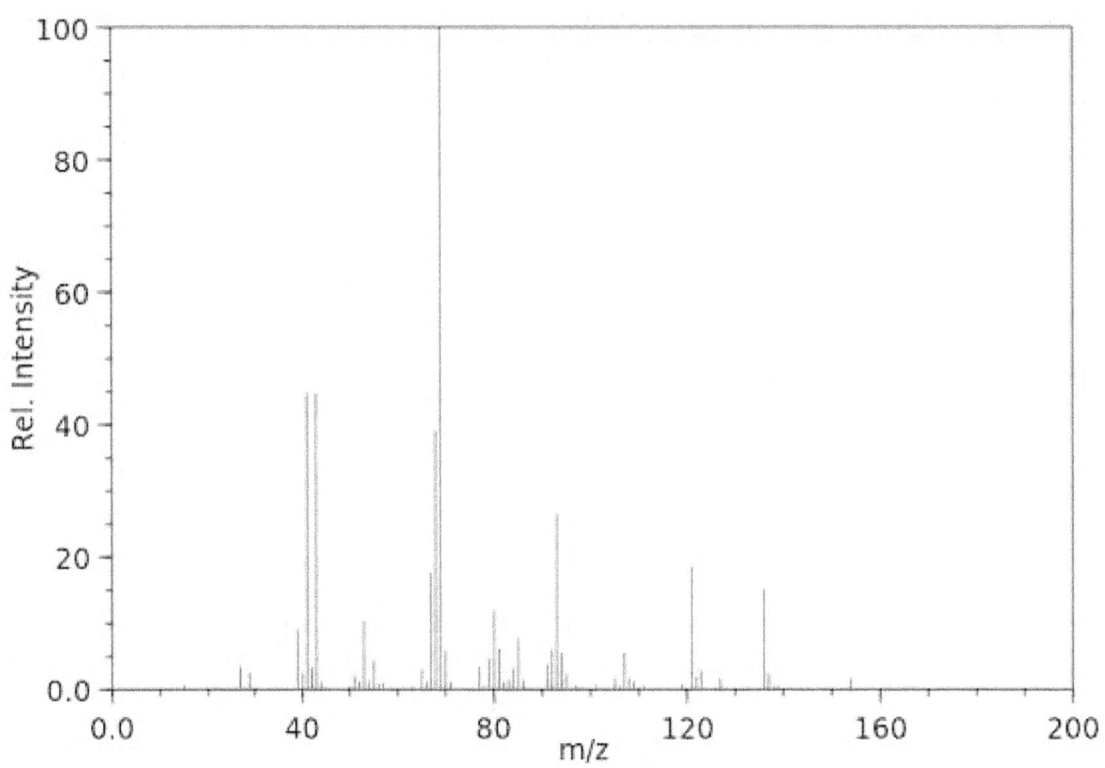

Observations

Charged fragments (m/z)	Assignment	Charged fragments (m/z)	Assignment
Molecular ion: 196	$[C_{12}H_{20}O_2]^+$	**Base peak: 69**	$[C_5H_9]^+$
154	$[C_{10}H_{18}O]^+$	43	$[C_3H_7]^+$
136	$[C_{10}H_{16}]^+$	39	$[C_3H_3]^+$
121	$[C_9H_{13}]^+$	29	$[C_2H_5]^+$
93	$[C_7H_9]^+$	15	$[CH_3]^+$

Conclusions

The mass spectrum confirms that the molecule is an ester and implies that it has a long, aliphatic, chain. It is neither possible to identify any substituted groups nor is it possible to identify the alkyl group forming part of the ester functional group and this is where the ^{1}H and ^{13}C nmr spectra become essential.

We consider these next.

NMR Spectra

^{1}H NMR Spectrum

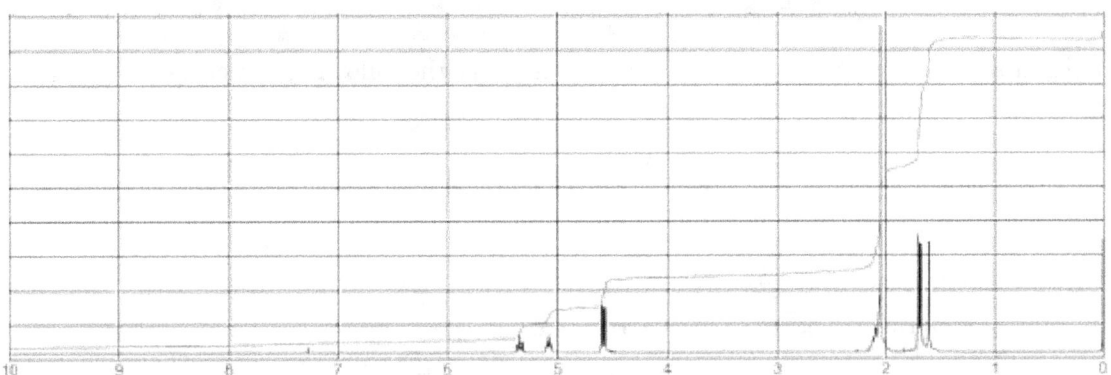

Compared to ethyl oleate, in the previous chapter, this spectrum is quite straightforward to analyse. We can observe the:-

- Peaks at δ 5.35 and δ 5.15 ppm, both of integral one, must be due to alkene hydrogen atoms;

- Doublet, of integral two, must be due to a $- OCH_2 -$ group;

- Singlet of integral six must be due to two methyl, $- CH_3 -$, groups attached to a carbon atom with no hydrogen atoms attached;

- Singlet of integral three at δ 1.6 ppm.

This means that there must be:-

- A $(- CH_3)_2 -$ group attached to a carbon atom with no other bonded hydrogen atom .

  That means that there must be a $(CH_3)_2 - C = C$ group in the molecule. One of the alkene hydrogen must be bonded to this group and both alkene hydrogen signals are triplets which means at least one end of the molecule must have the following structure:

where as usual, the squiggle ∿∿ represents an undetermined group.

The other end of the molecule can be considered next:

- A $- CH_3$ group attached to a carbon atom with no hydrogen atoms on the adjacent carbon atom must be the methyl group on the ester functional group.

- The $-O-CH_2-$ group due to the doublet, of integral two, at δ 4.6 ppm must be attached a carbon atom with only one hydrogen atom bonded to it (n+1 rule) and so it must be bonded to the other alkene group.

This means that this end of the molecule must have the following structure:

We can assemble the ends together as shown below with the squiggle representing the remaining part of the structure to be ascertained:

This accounts for eleven of the twelve carbon atoms and this last atom must be that in the methyl group which is a singlet of integral three.

- There must be a $-CH_3$ group attached to a carbon atom with no hydrogen atoms attached (*n+1 rule*)which accounts for the doublet of integral three.

This must be substituted on the chain.

There is only one place that the methyl group can exist, on the alkene $CH = C - R$ grouping as shown below:

as, although the alkyl group could be elsewhere in the molecule, it would not cause a doublet.

So putting this all together the only assembly of the functional groups can be the following structure:

There are two matters to consider: the coupling constants of the alkene group and the ^{13}C nmr spectrum. We consider the coupling constants before examining the ^{13}C nmr spectrum.

Geranyl acetate

Coupling Constants

We must consider the stereochemistry of the alkene groupings.

As discussed in the previous chapter, we can draw the molecule as either Z – (zusammen) or E – (entgegen) configurations as shown generalised below:

Z –   E –

where R and R' are any group but cannot be hydrogen atoms.

The data sheet records that for alkenes in general the, experimentally measured, coupling constants, are:-

- Zusammen (Z –) : J = 5 – 14 Hz
- Entgegen (E –) : J = 15 – 20 Hz

The coupling constants of the hydrogen atoms on both alkene bonds in this molecule are both J = 7 Hz and so, both bonds have the zusammen (Z –) stereochemistry.

Our final matter is to predict and examine the ^{13}C nmr spectrum.

^{13}C NMR Spectrum

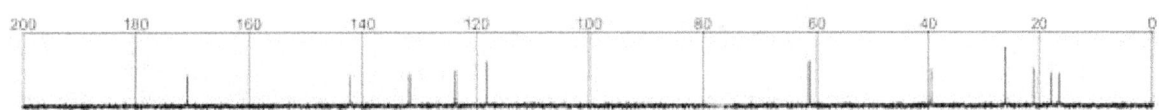

- The most important signal in the ^{13}C nmr spectrum is that at δ 26 ppm which is of integral two and represents the, two, chemically and magnetically equivalent carbon atoms in the methyl groups.

- There are four alkene carbon signals in the δ 100 – 150 ppm region.

- The peak, of integral one, at δ 173 ppm is assignable to the ester carbon atom.

This supports the structure as previously determined.

Geranyl acetate

Conclusions

Structure:

Systematic name: (Z –) – 3, (Z –) 7-Dimethylocta-2,6-dien-1-yl ethanoate.

Unless one knows the IUPAC naming rules in detail (which few people do without referring to the rules) this is one example where a simpler name is used more commonly.

Chapter XVI

Gossyplure

Gossyplure is a sex attractant employed by both pink bollworms (pictured above) and butterflies. It is used as a natural pesticide in organic cotton farming since it confuses male pink bollworms and disrupts reproduction. First found in the cotton belt of the southern United States in 1920, its eradication in that area was announced in 2018.

As is clear from the image above, the pink bollworm is an insect not a worm. Native to Asia, the adult is a small, thin, grey moth with fringed wings whilst the larva is a dull white caterpillar with eight pairs of legs with pink bands along its dorsum. The female pink bollworm moth lays eggs in a cotton boll – cotton bolls are the round, fluffy clumps in which cotton grows. When the larvae emerge from the eggs, they cause their damage by chewing through the cotton lint to feast on the seeds.

Effectively insoluble in water, gossyplure is a light yellow, odourless oil which has a melting point of 15°C and boiling point of 170°C.

This compound has the elemental composition: C: 77.02%, H: 11.52%; O: 11.41% and has the formula mass (M_r) of 280.45 g mol^{-1}.

This means that the formulas are as follows:

Empirical formula: $C_9H_{16}O$
Molecular formula: $C_{18}H_{32}O_2$.

Given the large number of carbon atoms this is either an aliphatic long, or substituted chain or aromatic compound. Since it contains two oxygen atoms it is likely that this molecule is a carboxylic acid or an ester.

Infrared Spectrum

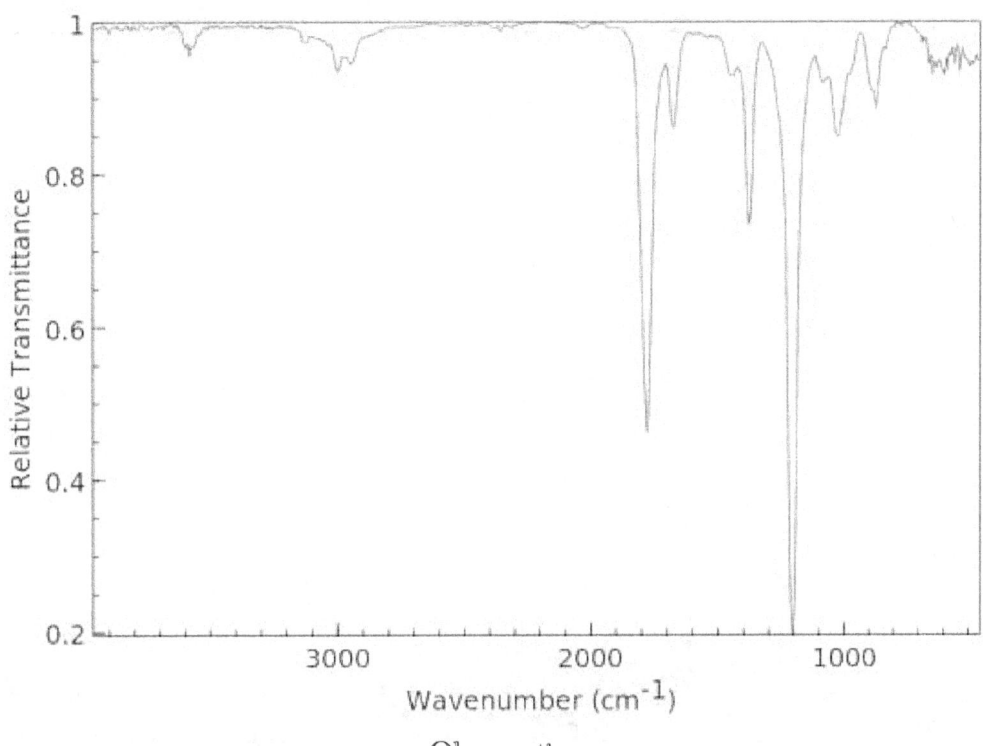

Observations

(√ / X)	Wavenumber range (cm⁻¹)	Wavenumber (cm⁻¹)	Assignment
X	3200 - 3700		O – H
X	3200 - 3600		N – H
X	3000 – 3300		C – H (aromatic)
√	2500 – 3000	2980,2920	C – H (aliphatic)
X	2200 – 2500		C ≡ N
√	1700 – 1800	1760	C = O
√	1600 – 1700	1620	C = C (aliphatic)
X	1585 – 1600		C – C (aromatic)
X	1450 – 1600		C – C (aromatic)
√	1000 – 1300	1200	C – O
X	700 – 1000		C – X (X = Cl, Br or I)

Conclusions

The formula already implies that this molecule is an ester, a carboxylic acid or a diol (a molecule with two – OH groups). The infra red spectrum shows the presence of a C = O and a C – O bond and an absence of an O – H stretch and so this molecule is clearly an ester. The short but sharp peak at 1620 cm⁻¹ also suggests the presence of at least one C = C bond.

Gossyplure

Mass Spectrum

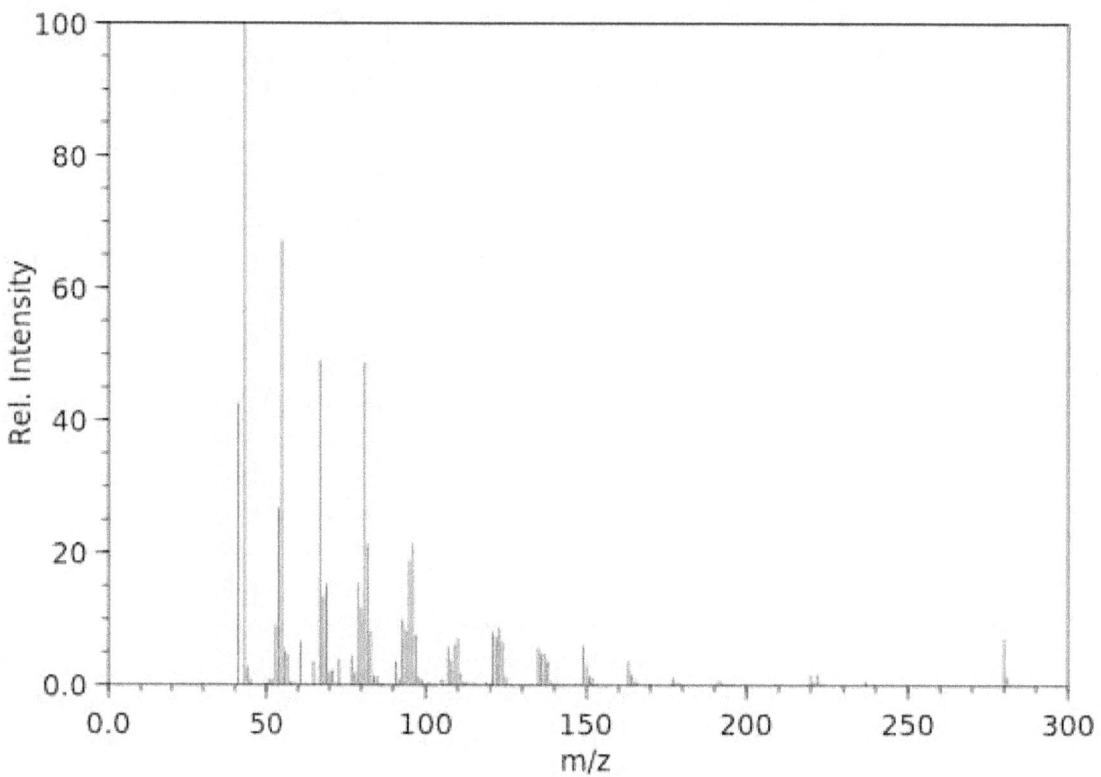

Observations

Charged fragments (m/z)	Assignment	Charged fragments (m/z)	Assignment
Molecular ion: 280	$[C_{18}H_{32}O_2]^+$	**Base peak: 43**	$[C_3H_7]^+$
237	$[C_{15}H_{25}O_2]^+$	110	$[C_8H_{14}]^+$
221	$[C_{15}H_{25}O]^+$	96	$[C_7H_{12}]^+$
149	$[C_{10}H_{13}O]^+$	81	$[C_6H_9]^+$
123	$[C_9H_{15}]^+$	67	$[C_6H_7]^+$

Conclusions

This spectrum indicates that the molecule has a long, aliphatic, chain or a shorter, more highly substituted chain and it does support the suggestion that it is an ester.

We can conclude the exact structure from the nmr spectrum and we consider the ¹H nmr spectrum next.

NMR Spectra

^{1}H NMR Spectrum

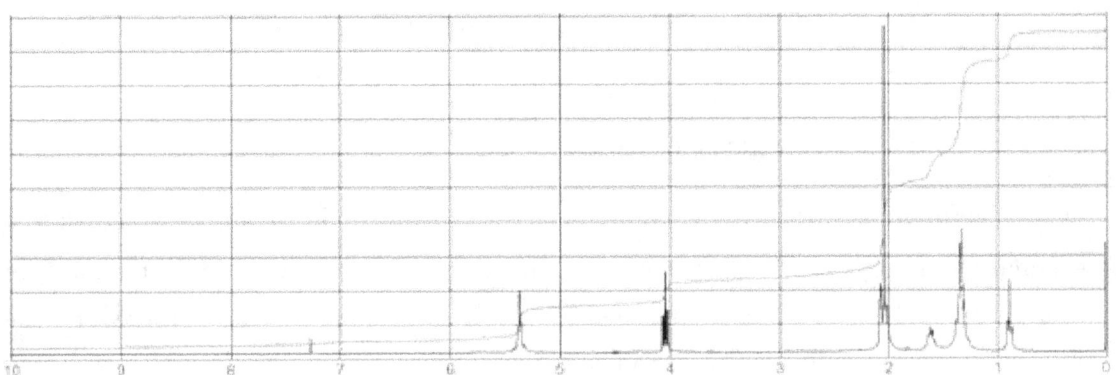

Observations

Chemical shift δ (ppm)	Multiplicity	Integral	Assignment
5.4	Doublet of triplets	4	– C(H) = C(H) –
4.05	Triplet	2	– OCH_2 –
2.1	Quintet	10	– CH2 –
1.65	Singlet	3	CH_3 – O –
1.35	Quintet	10	– CH2 –
0.9	Triplet	3	– CH_3

Conclusions

Some of these peaks are confusing but some are also straightforward to dispose of:-

 * The doublet of triplets, of integral four, at δ 5.4 ppm indicates that there must be two C = C bonds. Each of the double bonds must have a – CH_2 – bond attached to each carbon atom so we must find that the molecule contains the following grouping:-

 * The singlet, of integral three, at δ 1.65 ppm indicates that there is a single methyl (– CH3), group within the ester functional group.

 * The triplet, of integral two, at δ 4.05 ppm indicate that the – OCH_2 – group is attached to a – CH_2 – grouping.

This means that we can assign the peaks to the following fragments:

This leaves us with four more groups to assign.

■ The triplet, of integral three, at δ 0.9 ppm is assignable to, and can only be assigned to a terminal methyl group. The fact, however, that the signal is a triplet indicates that the terminal methyl group must be attached to a $-CH_2-$ group and so we can extend the right hand fragment above to be as shown below:

This leaves us with two $-CH_2-$ groups and they can only be used to link the two fragments together.

The only possible conclusion is that the structure of the molecule is:

With regard to the ^1H nmr spectrum the last matter to consider are the coupling constants of the C = C hydrogen atoms.

In both cases the coupling constants are J = 16 Hz and so both alkene groupings must be the E – stereoisomer. Remember that E – stands for the German term **entgegen** (*against*) and in earlier times these stereoisomers would have been described as *trans –*.

This means that the structure of the molecule can only be:

but we can confirm it by examining the ^{13}C nmr spectrum.

Gossyplure

¹³C NMR Spectrum

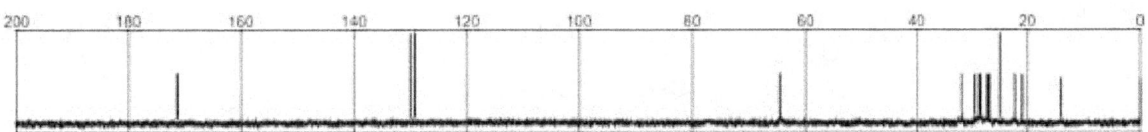

If we examine the ¹³C nmr spectrum in more detail we observe the following:

- One peak of integral one at δ 175 ppm due to the C = O carbon atom;
- Two peaks each of integral two at:
    - One peak, of integral two, at δ 131 ppm due to one of the pairs of the C = C carbon atoms;
    - A second peak, also of integral two, at δ 129 ppm due to the other pair of the C = C carbon atoms;
- A peak at δ 63 ppm which can be assigned to a C − O carbon atom.
- The peak, of integral two, at δ 24 ppm must be due to the − CH₂ − bond between the two C = C bonds;
- There is a peak at δ 15 ppm which can be assigned to the terminal methyl, − CH₃, carbon atom.

All of this is consistent with the proposed structure and there are nine peaks in the region δ 21 − δ 36 ppm which must be due to the rest of the carbon atoms and they must form part of a chain.

The peaks can, confidently, be assigned as follows:

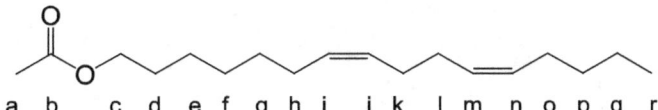

- C(a): δ 63 ppm;
- C(b): δ 21 ppm;
- C(c): δ 175 ppm;
- C(i) / C(j) & C(m) / C(n): δ 131 and δ 129 ppm;
- C(r): δ 15 ppm

The remainder of the peaks match the concept of the chain, aliphatic, ester.

Conclusions

Structure:

Systematic name: (7Z,11Z)-hexadeca-7,11-dien-1-yl ethanoate

Chapter XVII

Tiglal

Tiglal is a pheromone produced by female rabbits to attract their young to feed.
It has a melting point of -76ºC and a boiling point of 116ºC.

Tiglal has the following properties:
Elemental composition: C: 71.32%, H: 9.61%, O: 19.02%
Formula mass (M_r): of 184.12 g mol^{-1}.

This means that the empirical and molecular formulas are both C_5H_8O.

Compared to some of the other molecular structures analysed and determined in this
volume, this molecule is a delight to analyse.

Tiglal

Infrared Spectrum

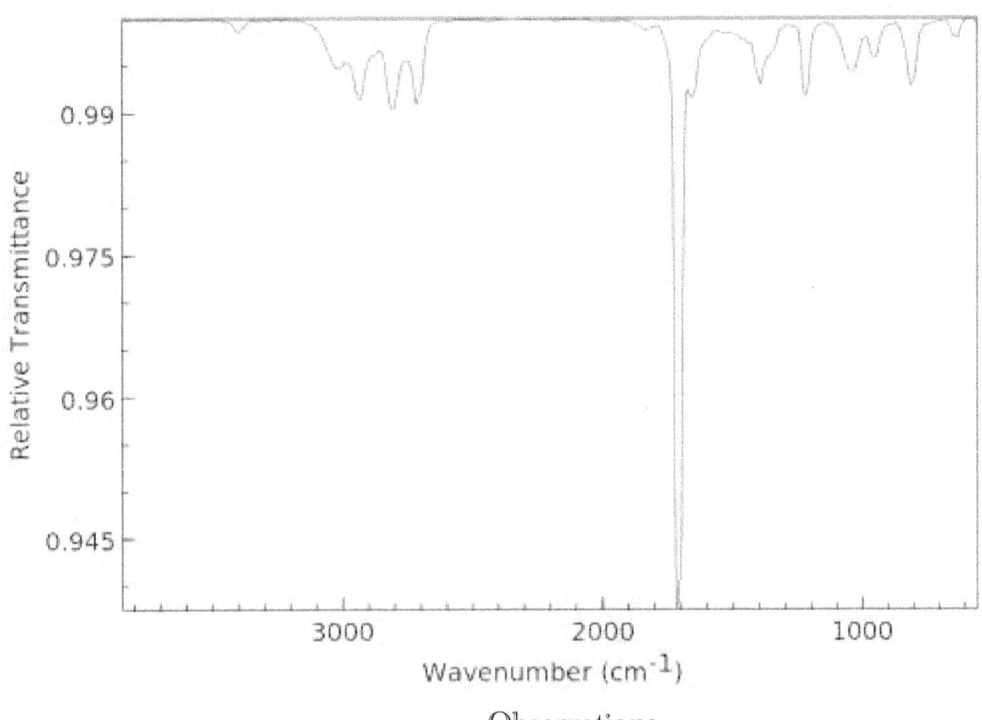

Observations

(√ / X)	Wavenumber range (cm⁻¹)	Wavenumber (cm⁻¹)	Assignment
√	3200 - 3700	3400	O – H
X	3200 - 3600		N – H
X	3000 – 3300		C – H (aromatic)
√	2500 – 3000	3020, 2920, 2820, 2650	C – H (aliphatic)
X	2200 – 2500		C ≡ N
√	1700 – 1800	1710	C = O
X	1600 – 1700		C = C (aliphatic)
X	1585 – 1600		C – C (aromatic)
X	1450 – 1600		C – C (aromatic)
√	1000 – 1300	1150	C – O
X	700 – 1000		C – X (X = Cl, Br or I)

Conclusions

This molecule is clearly an ester or an aldehyde and it is also aliphatic. There is a very small peak at 3400 cm⁻¹ which might indicate the presence of an O–H bond but it is not of characteristic shape or strength for a conventional –OH peak.

Tiglal

Mass Spectrum

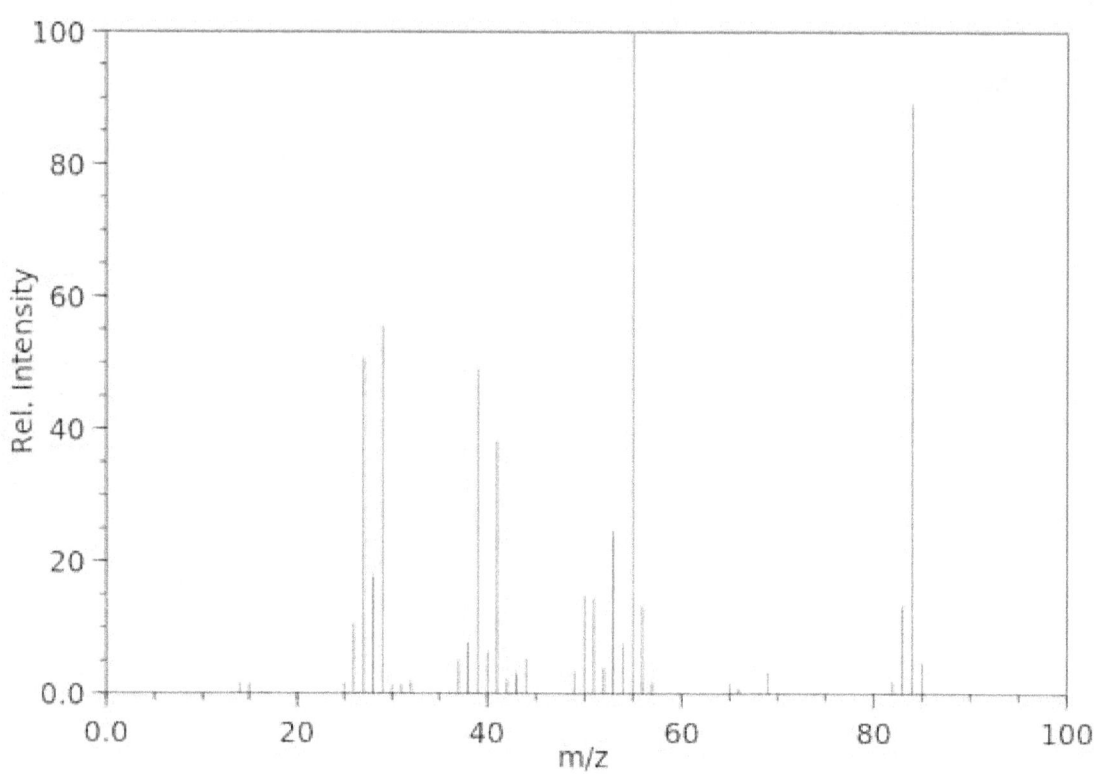

Observations

Charged fragments (m/z)	Assignment	Charged fragments (m/z)	Assignment
Molecular ion: 84	$[C_5H_8O]^+$	**Base peak: 55**	$[C_4H_7]^+$
69	$[C_5H_9]^+$	39	$[C_3H_3]^+$
41	$[C_3H_5]^+$ or $[C_2HO]^+$	29	$[C_2H_5]^+$ or $[CHO]^+$

Conclusions

There is not much to be concluded from this spectrum since, as the molecule is small, there are not many possible fragmentations. In addition, the two possible assignments for the peaks at m/z = 29 and 41 demonstrate that the assignments in the data charts are not absolute and are merely a useful guide.

The peak at m/z = 41 does suggest, however, that there is a $- OC=C(H) -$ group.

We can establish the exact structure from the ^1H and ^{13}C nmr spectra.

NMR Spectra

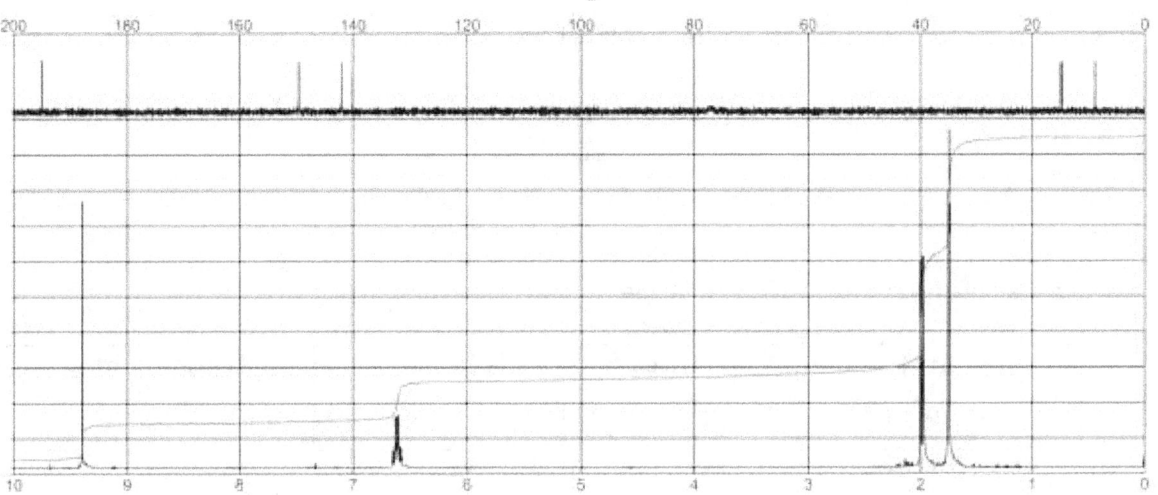

The ^{13}C nmr spectrum is easy to analyse so we will consider it first and can tabulate it:

Chemical shift δ (ppm)	Integral	Assignment(s)
196	1	–CHO
148 & 142	1 & 1	C = C
16 & 10	1 & 1	Alkyl

This means that there are very few possible structures.

▪ Aldehyde groups can only exist as a terminal functional group since the aldehyde carbon has three bonds already committed to the C = O bond and the C – H bond. If this group was within a chain then the carbon atom must bond to another *two* carbon atoms which is impossible.

▪ The presence of two alkene carbon atoms, both of integral one, indicates the presence of one C = C bond. This alkene bond cannot contain the aldehyde carbon atom as, again, it would break the valency rules for carbon since that carbon atom would have:-

  ▪ Two bonds to the oxygen atom,

  ▪ One bond to the hydrogen atom, and

  ▪ Two bonds to the other carbon atom in the C = C bond.

That is impossible and so the structure must be

which can be confirmed by examining the ^1H nmr spectrum which is our last task.

Tiglal

^{1}H NMR Spectrum

If we examine the proposed structure and label the hydrogen atoms

we can predict the following signals in the ^{1}H nmr spectrum:

Chemical shift δ (ppm)	Integral	Multiplicity	Assignment(s)
9 – 10	1	Singlet	H(a)
0.5 – 2	3	Singlet	H(b)
4.5 – 6	1	Quartet	H(c)
0.5 – 2	3	Doublet	H(d)

Observations and Conclusion

If we examine the ^{1}H nmr spectrum we can immediately identify the following multiplets:

Chemical shift δ (ppm)	Integral	Multiplicity	Assignment(s)
9.45	1	Singlet	H(a)
1.75	3	Singlet	H(b)
6.6	1	Quartet	H(c)
1.98	3	Doublet	H(d)

The quartet at δ 6.6 Hz has a coupling constant of 17 Hz so it is the entgegen (E –) isomer.

Conclusions

Structure:

Systematic name: (E)-2-methylbut-2-enal.

Nasonov Pheromones

Nasonov pheromones are very important substances which are released by worker bees to direct forager bees. It is believed that the pheromones are released by a bee raising its abdomen and beating its wings to fan the pheromones into the air after it has been released from the *Nasonov Gland*.

The Nasonov secretion gland occurs only in worker bees and are not found in Queen Bees or drone bees.

Nasonov pheromones perform a number of roles including:

- Directing young worker bees returning from orientation flights;
- Directing forager bees to a food source;
- Formation of a swarm of bees;
- Guiding disoriented returning forager bees;
- Instructing worker bees to await the return of the queen bee from a mating flight;
- To guide worker bees at queenless colonies.

To date, a number of Nasonov pheromones are have been identified and characterised:

- geraniol
- nerolic acid
- gerianal
- citronellate
- citral
- geranic acid
- nerol
- farnesol

Although each pheromone has some limited effect it has been found that the effect is greatest when all seven pheromones are present. It has also been found that the composition of the pheromone mixture varies with the season and age of the bee. We analyse three of these pheromones next.

Chapter XVIII

Geraniol

Geraniol is the first of the *Nasonov pheromones* to be considered in this volume.

It is produced by the scent glands of honey bees which marks nectar-bearing flowers and to also identify the entrances to hives. Bees also used it to repel other insects such as mosquitoes.

Essentially insoluble in water, geraniol is a colourless liquid with a rose-like odour. It is found in geraniums and lemons and is also detected in many essential oils such as rose oil and citronella oil. In the food industry, geraniol is used as a component in many artificial flavourings including those of peach, grapefruit, red apple, orange, watermelon and pineapple.

With a melting point of -15°C and a boiling point of 230°C, geraniol is a colourless liquid with the following properties:

Elemental composition: C: 77.77%, H: 11.78%, O: 10.37%
Formula mass (M_r): of 154.3 g mol^{-1}.

These means that the empirical and molecular formulas are both $C_{10}H_{18}O$.

Geraniol

Infrared Spectrum

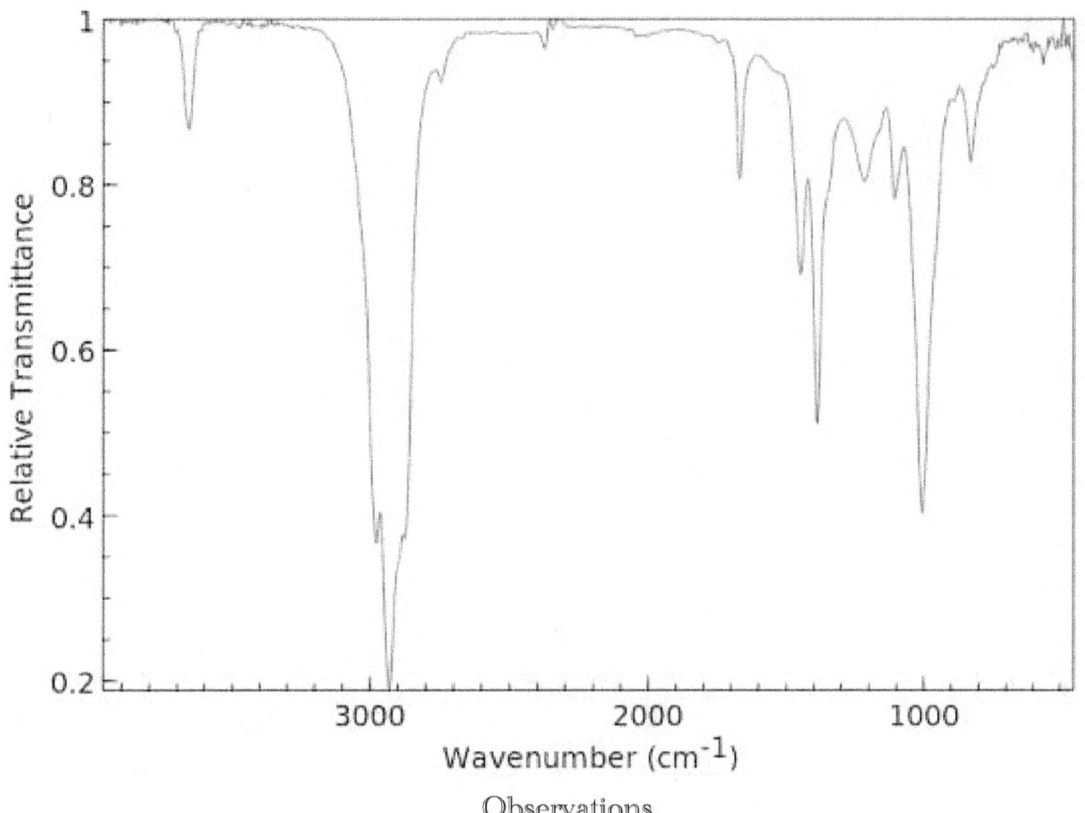

Observations

(√ / X)	Wavenumber range (cm⁻¹)	Wavenumber (cm⁻¹)	Assignment
√	3200 - 3700	3600	**O – H**
X	3200 - 3600		**N – H**
X	3000 – 3300		**C – H (aromatic)**
√	2500 – 3000	2980, 2930, 2860	**C – H (aliphatic)**
X	2200 – 2500		**C ≡ N**
X	1700 – 1800		**C = O**
X	1600 – 1700	1620	**C = C (aliphatic)**
X	1585 – 1600		**C – C (aromatic)**
X	1450 – 1600		**C – C (aromatic)**
√	1000 – 1300	1380	**C – O**
X	700 – 1000		**C – X (X = Cl, Br or I)**

Conclusions

This molecule is clearly a long chain aliphatic alcohol with at least one C = C bond but we can learn more from the mass spectrum and then from the ^{1}H and ^{13}C nmr spectra.

Geraniol

Mass Spectrum

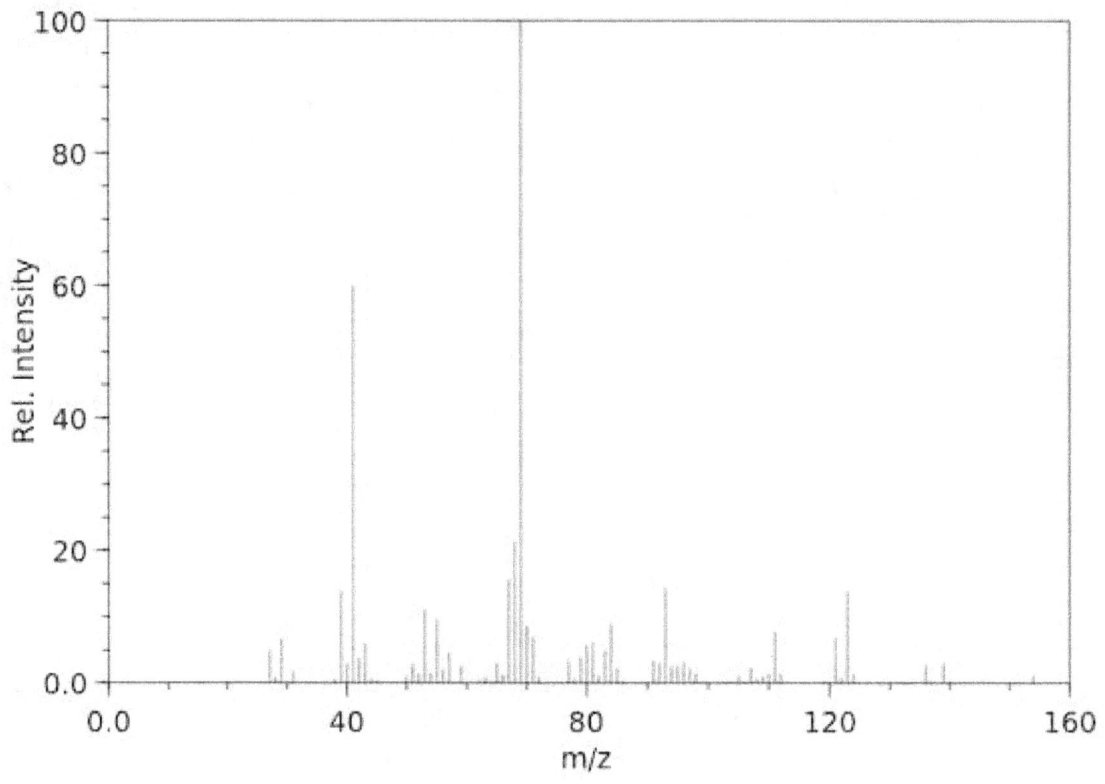

Observations

Charged fragments (m/z)	Assignment	Charged fragments (m/z)	Assignment
Molecular ion: 154	$[C_{10}H_{18}O]^+$	**Base peak: 69**	$[C_5H_9]^+$
139	$[C_9H_{15}O]^+$	39	$[C_3H_7]^+$
93	$[C_7H_9]^+$	29	$[C_2H_5]^+$
41	$[C_3H_5]^+$	15	$[CH_3]^+$

Conclusions

This spectrum implies that the molecule is a long chain alcohol. There must be at least one C = C bond however the molecular formula implies that there are two C = C bonds as there are insufficient hydrogen atoms for there to be one only.

We can lean more from the nmr spectra and we consider the ¹H nmr spectrum next.

Geraniol

NMR Spectra

¹H NMR Spectrum

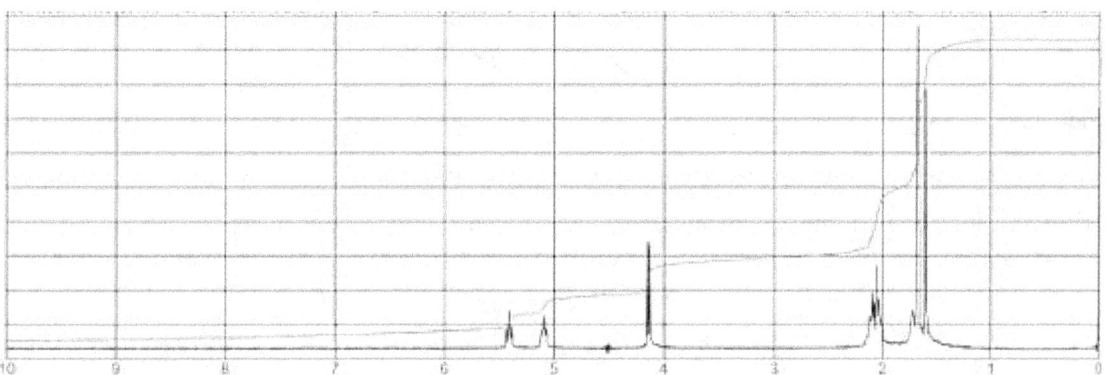

There are clearly six distinct chemically and magnetically distinct environments and we can immediately dispose of some as discussed below:

- The triplets at δ 5.45 ppm and δ 5.10 ppm, each of integral one must be due to alkene hydrogen atoms and this indicates the presence of two alkene groups. Since each is of integral one then each of the $C = C$ groups must have only one hydrogen atom attached. This means that they must be somewhere within the molecule each with three carbon atoms attached. The fact that they are both triplets is highly significant and we discuss that later on this page.

- The singlet of integral six at δ 1.6 ppm indicates the presence of a pair of methyl groups attached to a carbon atom. That the signal is a singlet indicates that they are both bonded to a carbon atom with no hydrogen atoms attached. The only explanation is the following structure:

where the squiggle, ~~~ , represents an undetermined group.

- Each of the alkene hydrogen atoms produces a signal which is a triplet and this means that we can replace the squiggle with a $- CH_2 -$ group so the molecular fragment now becomes

The coupling constants for both alkene groups are $J = 17$ Hz so both alkene groups have the entgegen (E −) stereochemistry.

Geraniol

- The $-CH_2-$ group, of integral two, produces a doublet of triplets and this means that it must have another $-CH_2-$ group attached to it so the molecular fragment now becomes:

- There is an alkyl triplet of integral one and this must be attached to the carbon atom before the squiggle but must also be bonded to the other $C = C$ group. It also means that that there must be a methyl group attached and so the fragment becomes:

- There is one more $-CH_2-$ and an $-OH$ group to place on the structure and so the molecule must have the following structure:

It must be remembered that the $-OH$ hydrogen often does not appear in the ^1H nmr spectrum, due to hydrogen exchange with the solvent, but the infrared spectrum and the molecular formula indicates its presence.

The structure is confirmed by the ^{13}C nmr spectrum:

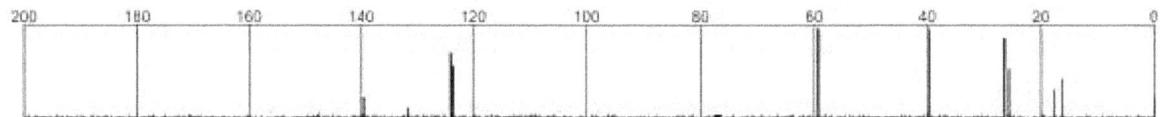

which confirms the presence of four alkene carbon atoms, a $C - OH$ group and five alkyl carbon atoms of which the most significant is the singlet, of integral two, at δ 23 ppm which is caused by the two chemically and magnetically equivalent methyl groups.

Conclusions

Structure:

Systematic name: $(E) - 3,(E -) - 7 -$ dimethylocta-2,6-dien-1-ol.

This molecule is isomeric with nerol (chapter XIII)

Chapter XIX

Citral

Citral, the second of the Nasonov pheromones to be considered in this text is a very pale yellow oil with a melting point of -10°C and a boiling point of 229°C.

Citral is found in many plants including lemon balm, lemon, lime, orange and the lemon tea-tree as well as many species of trees including the eucalyptus, Japanese cypress (pictured above) and the Nepal paper plant. It is detected by forager bees to identify pollen sources.

This compound has the elemental composition: C: 78.83%, H: 10.61%, O: 10.51% and has the formula mass (M_r) of 152.24 g mol^{-1}.

This means that it has the following formulas:

Empirical formula: $C_{10}H_{16}O$
Molecular formula: $C_{10}H_{16}O$.

Infrared Spectrum

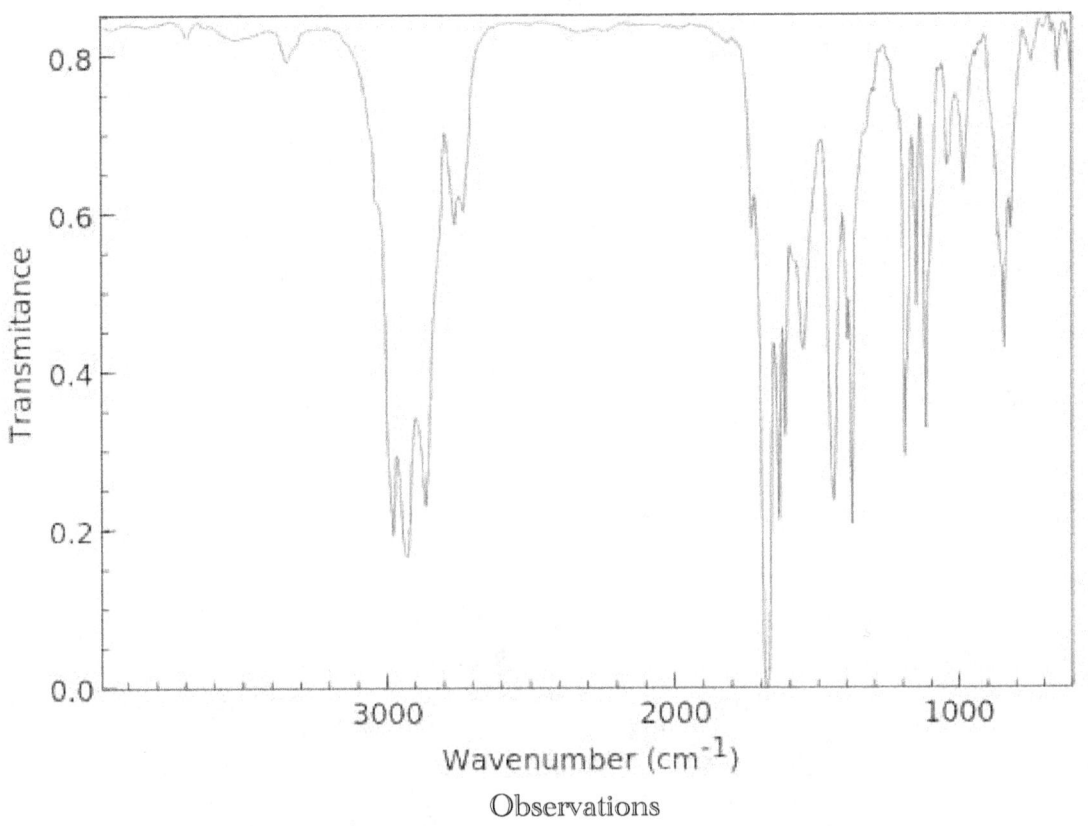

Observations

(√ / X)	Wavenumber range (cm⁻¹)	Wavenumber (cm⁻¹)	Assignment
X	3200 - 3700		O – H
X	3200 - 3600		N – H
X	3000 – 3300		C – H (aromatic)
√	2500 – 3000	2980, 2920, 2840, 2780, 2730	C – H (aliphatic)
X	2200 – 2500		C ≡ N
√	1700 – 1800	1680	C = O
√	1600 – 1700	1620	C = C (aliphatic)
X	1585 – 1600		C – C (aromatic)
X	1450 – 1600		C – C (aromatic)
√	1000 – 1300	1160	C – O
X	700 – 1000		C – X (X = Cl, Br or I)

Conclusions

This molecule contains one oxygen atom and so cannot be an ester. It cannot be an alcohol as there is no O – H stretch. The presence of the C = O and C = C peaks indicates it must be an unsaturated aldehyde or ketone.

Citral

Mass Spectrum

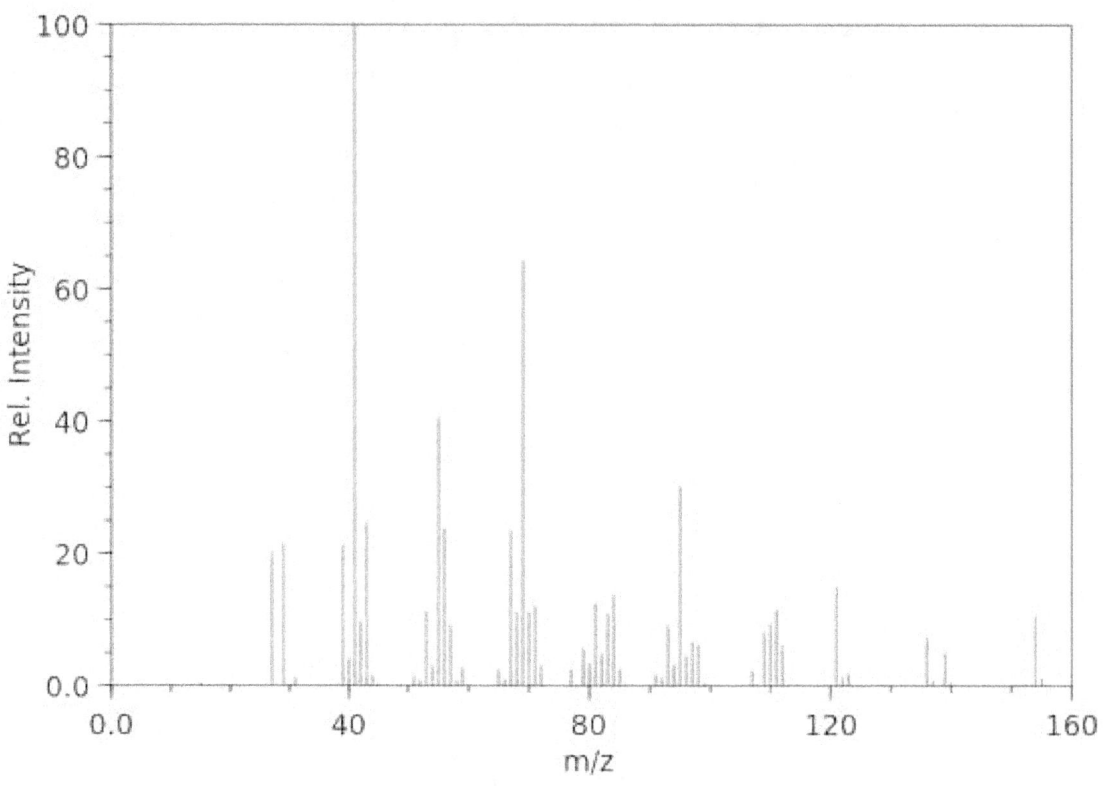

Observations

Charged fragments (m/z)	Assignment	Charged fragments (m/z)	Assignment
Molecular ion: 152	$[C_{10}H_{16}O]^+$	Base peak: 69	$[C_5H_9]^+$
137	$[C_9H_{13}O]^+$	84	$[C_6H_{12}]^+$
123	$[C_8H_{11}O]^+$	41	$[C_3H_5]^+$
109	$[C_7H_9O]^+$	39	$[C_3H_3]^+$
94	$[C_6H_6O]^+$	29	$[C_2H_5]^+$

Conclusions

The mass spectrum supports the idea that the molecule is a long chain, aliphatic, compound but does not provide too much more information and so we must consider the ^{1}H and ^{13}C nmr spectra which is our next task.

We will consider the ^{1}H nmr spectrum first.

NMR Spectra

^{1}H NMR Spectrum

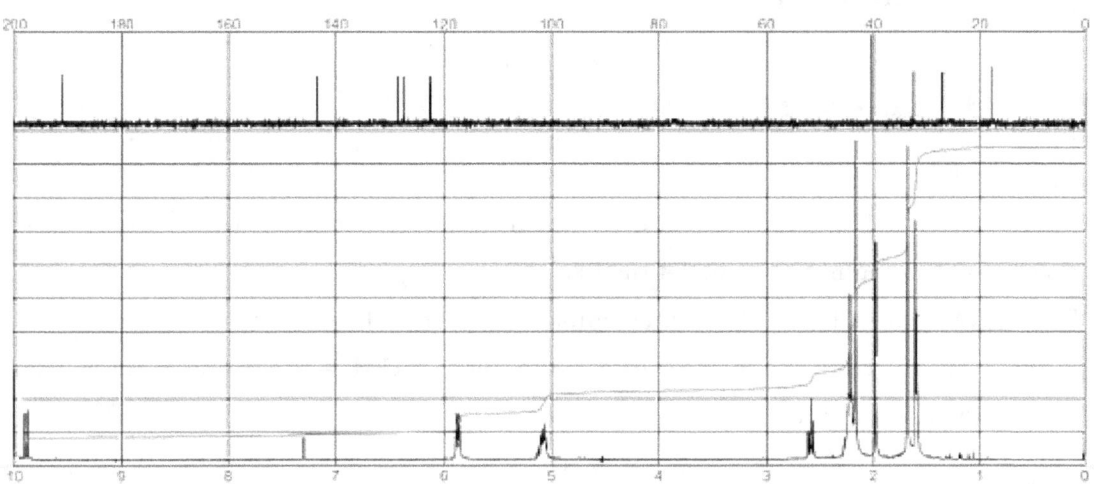

We can, very readily, assign and dispose of some of these peaks:

- The peak at δ 7.3 ppm is due to contamination of the CDCl$_3$ solvent and can be ignored.

- The doublet at δ 9.9 ppm indicates that the molecule is an aldehyde but the – CHO hydrogen atom is attached to a carbon atom with a single hydrogen atom attached. This means that the – CHO group is attached to a C(H) = CH$_2$ – group.

- The doublet of integral one at δ 5.9 ppm indicates the presence of a single hydrogen atom attached to immediately adjacent carbon atoms and so we can assign an initial fragment as shown below:

where the squiggle, ∿ , indicates an undetermined group.

- There is another alkene hydrogen signal, also of integral one, which is a triplet. This must mean that that the alkene hydrogen atom must have a – CH$_2$ – attached adjacently and so we also have this fragment:

◼ There is also a singlet, of integral six, which must must be due to two methyl, – CH₃, groups bonded to a carbon atom which has no other hydrogen atoms attached and so there must also be this grouping:

◼ There is an alkyl singlet of integral three which must be a substituent on to the chain otherwise it would not be a singlet (n+1 rule).

◼ This leaves two – CH₂ – CH₂ – .which must form part of the long carbon chain.

    ◼ One is a triplet so must be attached to the C = C(H) – group and so we now have the following sub-structure:

    ◼ The other – CH₂ – group is a doublet of triplets which means that it must be bonded to one – CH₂ – (causing the triplet) and also a – CH = group causing a doublet of triplets.

Putting this all together the only possible structure is as shown below:

There is one matter left to address and that is the orientation of the hydrogen atoms on the highlighted alkene groups.

We can determine the stereochemistry of this group by examining the coupling constant, J, of the highlighted groups. The coupling constants for the highlighted functional groups are both J = 7 Hz so in both cases the molecule must be the E – (entgegen) stereoisomer.

The remaining task is to examine the ^{13}C nmr spectrum in order to prove or disprove the conclusions.

Citral

^{13}C NMR Spectrum

If we consider the assumed structure again,

we can expect to observe the following:

- One peak, of integral one, in the range δ 160 – 220 ppm due to the C = O carbon atom;

- Four peaks, each of integral one, somewhere between δ 110 ppm and δ 160 ppm due to the, highlighted, C = C carbon atoms.

- Four peaks between δ 0 and 50 ppm.

  - Two peaks, each of integral one, will be due to the, so far, unassigned carbon atoms in the chain;

  - One peak, of integral one, will be due to the carbon atom of the methyl, – CH$_3$ – substituent carbon;

  - One peak of integral **two**, will be due to the carbon atoms in the two chemically and magnetically equivalent carbon atoms.

Citral

This is exactly what we observe:

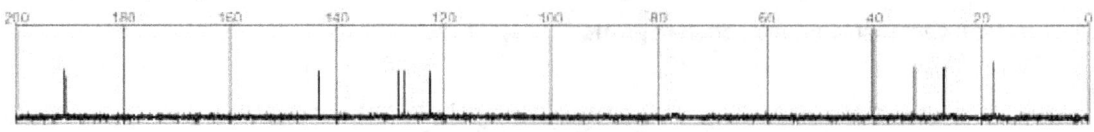

Conclusions

Structure:

Systematic name: (E)-3,7-dimethylocta-2,6-dienal

Chapter XX

Farnesol

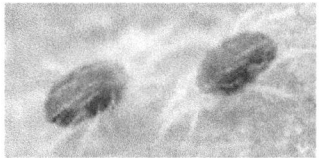

A pale yellow oil with a floral odour, **farnesol** is an important precursor for many natural products which is present in many essential oils. It is another pheromone used by forager bumble bees to identify sources of pollen and as a sex attractant.

Farnesol is an intriguing compound as it also employed as an alarm pheromone by spider mites and as a sex attractant by dung beetles (pictured below).

Commercially it is also used as an additive in cigarettes and vaping devices whilst, with apparent anti-bacterial properties it is also used in many deodorants

With a melting point of -100°C and boiling point of 111°C, this compound has the **elemental composition**: C: 80.95%, H: 11.81%; O: 7.19% and a **formula mass (M_r)** of 222.4 g mol^{-1}.

This means that the empirical and molecular formulas are both $C_{15}H_{26}O$.

Farnesol

Infrared Spectrum

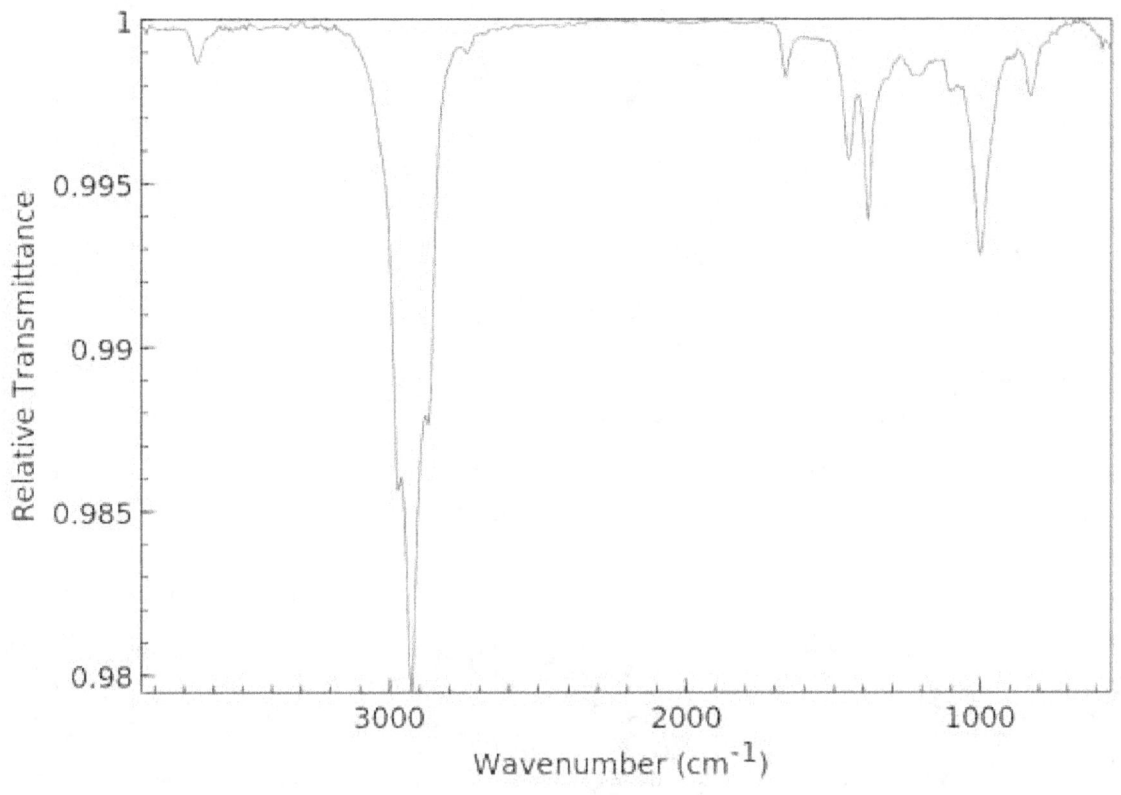

Observations

($\sqrt{}$ / X)	Wavenumber range (cm⁻¹)	Wavenumber (cm⁻¹)	Assignment
$\sqrt{}$	3200 - 3700	3660	O – H
X	3200 - 3600		N – H
X	3000 – 3300		C – H (aromatic)
$\sqrt{}$	2500 – 3000	2980, 2920, 2860	C – H (aliphatic)
X	2200 – 2500		C ≡ N
X	1700 – 1800		C = O
$\sqrt{}$	1600 – 1700	1620	C = C (aliphatic)
X	1585 – 1600		C – C (aromatic)
X	1450 – 1600		C – C (aromatic)
$\sqrt{}$	1000 – 1300	1005	C – O
X	700 – 1000		C – X (X = Cl, Br or I)

Conclusions

This molecule is clearly an aliphatic, unsaturated alcohol.

Farnesol

Mass Spectrum

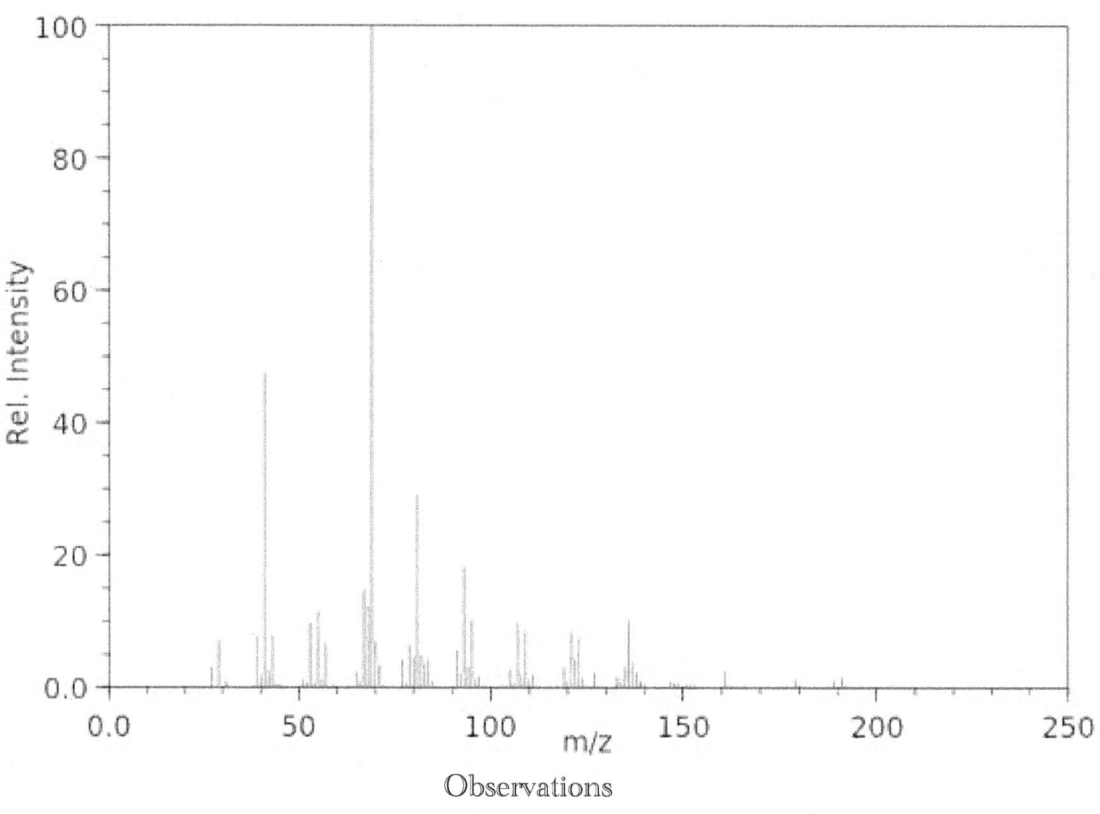

Observations

Charged fragments (m/z)	Assignment	Charged fragments (m/z)	Assignment
Molecular ion: 191	$[C_{14}H_{23}]^+$	**Base peak: 69**	$[C_5H_{59}]^+$

136	$[C_9H_{12}O]^+$	93	$[C_7H_9]^+$
123	$[C_8H_{11}O]^+$	55	$[C_4H_7]^+$
109	$[C_7H_9O]^+$	41	$[C_3H_5]^+$

Conclusions

This spectrum support the suggestion that this molecule is a long chain alcohol.

It does not, however, enlighten us as to the extent of substitution on the chain and this means we must rely on the ^{1}H and ^{13}C nmr spectra which we consider next.

Farnesol

NMR Spectra

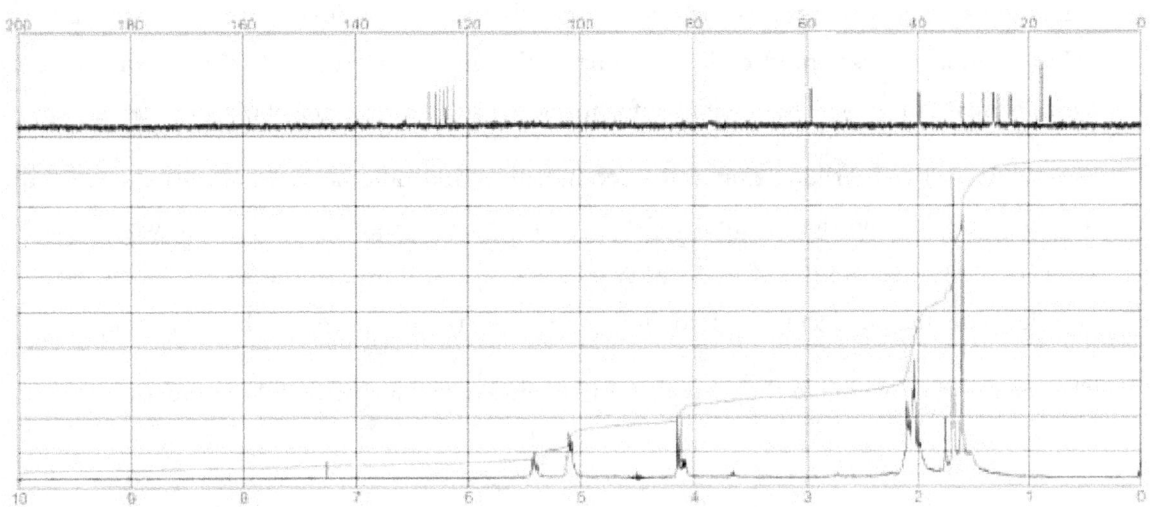

We will consider the ^{13}C nmr spectrum first.

^{13}C NMR Spectrum

- We can see immediately that there are six alkene carbon atoms:

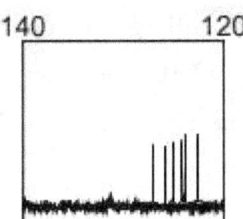

and this means that there must be three C = C bonds which is consistent with the molecular formula.

- There is a C – O group at δ 59 ppm. From the infrared spectrum we can see that there is an O – H group but no C = O peak and so the molecule must be an alcohol.

- Another task with the ^{13}C nmr spectrum is to examine the alkyl region which is shown below in expanded form:

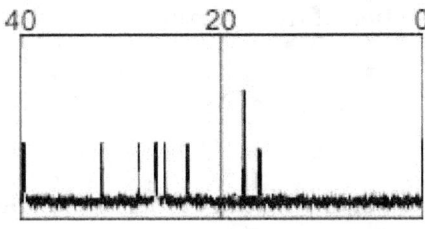

■ Of the eight peaks, the one of most interest is that at δ 18 ppm since it is of integral two. This indicates that there must be two chemically and magnetically equivalent methyl, –CH3, groups. These could be in any part of the molecule provided that the molecule is symmetrical but, given its molecular formula, the molecule cannot be symmetrical.

Since these two carbon atoms are chemically and magnetically equivalent in an assymetric molecule they cannot exist in two separate regions since there would be, by definition, another singlet of integral two. This is absent and so the straightforward and simplest conclusion is that the two methyl groups are on one end of the molecule.

This means that we have the structure of one end of the molecule which must be:

Since the molecular formula is $C_{15}H_{26}O$ this means that we have twelve remaining carbon atoms to assign and must include the three C = C bonds.

■ The next point though is that all of the alkyl carbon atoms are chemically and magnetically *non-equivalent* and that means the molecule cannot be symmetrical since the carbons in the symmetrical part would appear at the same chemical shift and they do not.

As a first guess, and all spectroscopists make guesses to either support or discard possible structures, we can suggest that the other terminal end of the molecule contains the alcoholic functional group.

This would then lead to, as a first guess, that the structure could be, generically, of the following form:

This means that the carbon atom bonded to the –OH group must be a – CH2 – group which lead us to the next interation of the structure:

We have now accounted for five of the fifteen carbon atoms but we need to resolve the position of the six carbon atoms which constitute the three C = C bonds and then position the remaining carbon atoms. Given that there are two further C = C bonds to position, the ^{13}C nmr spectrum indicates that some of the carbon atoms must be substituents on the chain as they would appear elsewhere if not substituents.

- The alkene carbon atoms are all completely and magnetically *non-equivalent* and this means that they must be bonded to carbon atoms which are themselves also chemically and magnetically non-equivalent.
- From the ^{1}H nmr spectrum, the OH – CH$_2$ – group must have a C = C bond attached as it is a singlet. This means that the left hand side of the molecule becomes:

Together, we then have the following, general, structure:

where the squiggle, $\sim\!\sim\!\sim$, represents the rest of the chain and we have nine more carbon atoms to fit into the structure.

There are six alkene carbon atoms but only three alkene hydrogen signals. This definitely shows that three of the alkene carbon atoms must only be bonded to other carbon atoms which shows that, if part of the chain, they must have methyl substituents attached.

It is worth noting that we are only referring to methyl substituents as, although it is possible to draw structures with ethyl, – C$_2$H$_5$, groups the alkyl region would include triplets and quartets and those signals are absent.

The methyl groups must be substituted on the chain and one of these must be on the C = C bond shown above and so we now have the following basic structure:

Given that there are another two C = C bonds to fit in which *must* be in the chain this again implies that the chain is not as long as possible but must be shorter and substituted. It is extremely unlikely that the remaining two C=C bonds will be immediately adjacent to each other and so each of the two C = C bonds must be separated by at least one methylene, – CH$_2$ – group.

Farnesol

This suggests that the structure must be of the general form:

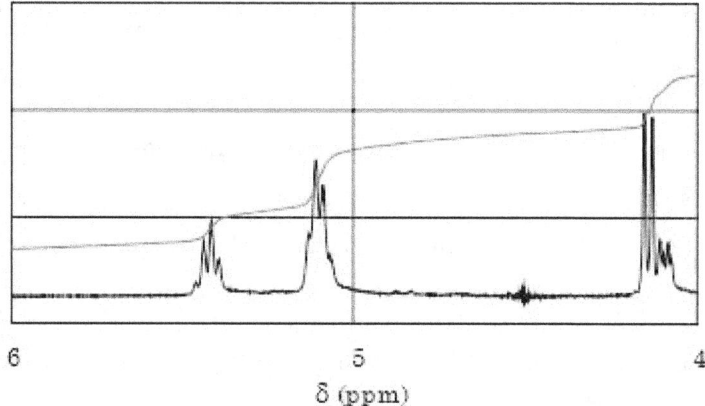

There are other isomers in terms of the C = C bonds being in slightly different positions and we also have fifteen hydrogen atoms to fit in which accords with the above proposition.

If we re-examine the alkyl region of the ^{13}C nmr spectrum then we can establish that there must be four methyl, $- CH_3$, substituents. We have accounted for two of the $- CH_3$ groups which form the terminal functional group. The other two must be somewhere on the chain. This now means we must consider the ^{1}H nmr spectrum which is our final task in this chapter and in this volume.

^{1}H NMR Spectrum

If we consider the, expanded, alkene region first we observe the following:

We observe:-

- A quintet of integral *one* at δ 5.45 ppm which indicates that this hydrogen atom must be bonded to *two* $- CH_2 -$ carbon atom adjacent to a C = C bond.
- A triplet of integral *two* at δ 5.15 ppm. This indicates that the signal results from two, completely separate hydrogen atoms which are, individually, bonded to a carbon atom which is bonded to a C = C carbon atom.
- A doublet of integral two at δ 4.2 ppm which must be due to a OH $- CH_2 - CH -$ group. This confirms the structure of one terminal grouping of the molecule.

Farnesol

The final task in this chapter and in this entire volume must be to examine the alkyl region which is shown, in expanded form, below:

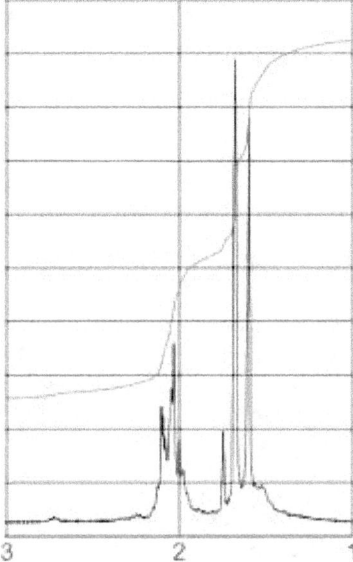

This portion of the spectrum confirms the existence of three groups of hydrogen atoms in the ratio 2:2:1 which accords with the structural formula and is 6:3:3.

This confirms the structure to be:

and our final task is to establish the stereochemistry of the alkene hydrogen atoms.

In all cases of the alkene hydrogen atoms the coupling constant is J = 17 Hz so, from the data sheet, we can conclude that all the stereochemistries are entgegen (E −).

Conclusions

Structure:

Systematic name: (2E,6E)-3,7,11-Trimethyldodeca-2,6,10-trien-1-ol

Part III

Indexes

This section is organised into several indexes with the contents organised by the:-

- Trivial name of the pheromone;
- Systematic name of the pheromone;
- The occurrence of pheromones in nature.

Index of Pheromones by Trivial Name

In this index, pheromones, indexed by name have the chapter number and the pages of the chapter listed. There is, of course, also the introductory cover page preceding each chapter but the cover pages are not numbered.

Index of Pheromones by Systematic Name

In this index, pheromones, are indexed by systematic name by the first letter of the name. the numbering and the E – or Z – nomenclatures are ignored for the purpose of indexing.

Since they are extremely unwieldy, the trivial name is included after the chapter number and the pages of the chapter listed. The page numbering does not include the cover page for that chapter.

Index of Pheromones by Occurrence

In this index, pheromones discussed in this volume are indexed by their currently known occurrence in nature. The list is inclusive of all pheromones and incidence described in this volume but it is far from a complete list of pheromones and their occurrence in nature.

Since the systematic names are extremely unwieldy the index includes the trivial names of the pheromones only as well as the pages of the chapter listed but, again, the page numbering does not include the cover page for that chapter

Millipedes	creosol	VI (37 – 46)
Oriental fruit moths	ethyl caproate	XI (79 – 84)
Pink bollworms	gossyplure	XVI (109 – 114)
Queen bee retinue	linolenic acid	IX (65 – 72)
Queen bee retinue	cetyl alcohol	XII (85 – 90)
Spider mites	farnesol	XX (129 – 135)
Stingless bees	nerol	XIII (91 – 96)
Termites	styron	VIII (53 – 64)
Wolves	methylparaben	II (9 – 18)
Yellow and black mud daubers	geranyl acetate	XV (103 – 108)

www.ingramcontent.com/pod-product-compliance
Lightning Source LLC
Chambersburg PA
CBHW080958170526
45158CB00010B/2833